PRIZE
FIGHT

PRIZE FIGHT

THE RACE AND THE RIVALRY

TO BE THE FIRST IN SCIENCE

Morton A. Meyers, MD

PRIZE FIGHT
Copyright © Morton A. Meyers, 2012.

All rights reserved.

First published in hardcover in 2012 by PALGRAVE MACMILLAN® in the US—a division of St. Martin's Press LLC, 175 Fifth Avenue, New York, NY 10010.

Where this book is distributed in the UK, Europe and the rest of the world, this is by Palgrave Macmillan, a division of Macmillan Publishers Limited, registered in England, company number 785998, of Houndmills, Basingstoke, Hampshire RG21 6XS.

Palgrave Macmillan is the global academic imprint of the above companies and has companies and representatives throughout the world.

Palgrave® and Macmillan® are registered trademarks in the United States, the United Kingdom, Europe and other countries.

ISBN 978–1–137–27842–5

Library of Congress Cataloging-in-Publication Data for the hardcover is below.

Meyers, Morton A.
    Prize fight : the race and the rivalry to be the first in science / Morton A. Meyers.
        p. cm.
    Includes bibliographical references and index.
    ISBN 978–0–230–33890–6 (hardback)
        1. Research—Moral and ethical aspects. 2. Scientists—Psychology. 3. Scientists—Professional relationships. I. Title.

Q180.55.M67M49 2012
174′.95—dc23   2011047900

A catalogue record of the book is available from the British Library.

Design by Newgen Knowledge Works (P) Ltd., Chennai, India.

First PALGRAVE MACMILLAN paperback edition: November 2013

10 9 8 7 6 5 4 3 2 1

Printed in the United States of America.

Dedicated to

Bea, Amy, Richard, Karen, Sarah, and Sam

Yes, Virginia, scientists do love recognition, but only since Pythagoras.

—*Leon Lederman, Nobel laureate*

CONTENTS

INTRODUCTION

"LOOK AWAY FROM THE BALL"

In basketball this rule highlights how the game is really played. The point guard of a college basketball team once asked his coach how he could improve his playing. The coach asked him what he did in practice. "Pass, dribble, and shoot," the player replied, indicating that he kept the ball in his control. The coach nodded his head and told him to have someone clock how much time he actually spent handling the ball in a regulation game. The player was surprised to find that he had his hands on the basketball less than three total minutes out of a forty-minute game.

"What do you learn from that?" asked his coach.

"Beats me," said the player.

"You learn," said the coach, "that most of the game is played away from the ball."[1]

A good game of basketball is not just exciting slam-dunks and three-point shots; it also involves competitive spirit, the race against the clock, fouls, time-outs, and, of course, overall strategy. The real challenge is not to focus on where the ball is at any particular moment but rather where it is likely to be in the next play. Sportswriters tend to burnish the image of a particular player without due acknowledgment of the contributions of the teammates or the aggressive actions of the opposing players.

So it is in science. The Lasker Awards, presented by the American Lasker Foundation, are announced at the end of September and often presage future recognition by the Nobel committee, so they have become popularly known as "America's Nobels." The Nobel Prizes, the jewel in the crown of scientific achievements, are announced in October each year. But it is not just the announcement and recognition of a scientific discovery that is meaningful. We must also "look away from the ball." We must seek to understand the continuing process, the enduring fundamentals, the richly textured human dramas swirling around pivotal discoveries that are as true and relevant today as they were yesterday. To focus solely or predominantly on the science distracts from the underlying human factors. The lens through which these human interactions can be meaningfully viewed is the domain of the sociology of science.

This book explores a fundamental question underlying scientific research. Whose work earns the hallmark of priority for an original meaningful discovery of a scientific truth? How is credit determined, allocated, and contested? The literature contains very little on the subject, despite the fact that it is a pervasive source of discomfort, and not uncommonly agony, among researchers. This is reflected in the acceptance speech by one of the winners of the 2004 Nobel Prize in physics, H. David Politzer, noteworthy for his unusually candid portrayal of the human interactions involved. The address is cogently titled "The Dilemma of Attribution." Politzer concludes: "More of the public should contemplate these matters [regarding the competitive drive for recognition] if they wish to understand not just the ideas of science but also how they have developed."[2]

Even in disparate academic circles, the issue of allocation and misallocation of credit has generally received scant notice. A controversy over proper attribution within one discipline hardly causes a ripple within another. Should it receive attention in the media, it is usually shrugged off as an aberration, a misadventure under a set of unique circumstances. But in 2003 the unprecedented public outburst by Raymond Damadian over his exclusion from being awarded the Noble Prize for magnetic resonance imaging (MRI) revealed the all-too-human passions behind the headlines. MRI was not an abstract advance of remotely theoretical potential but one of practical benefits familiar to many. Damadian's outburst was an appeal to the general public—MRI now being a household term—as well as the scientific community and the Nobel committee.

And then in October 2010, the long-simmering dispute over the priority of identification of the AIDS virus—which had required the personal intervention of the president of the United States and the prime minister of France—was officially resolved with the awarding of the Nobel Prize in medicine to Luc Montagnier of the Pasteur Institute in Paris. The controversy fomented aggressive behavior between Montagnier and his rival, Robert Gallo, a respected researcher at the National Cancer Institute, and attracted the attention of scientists, physicians, AIDS activists, government officials, and the general public.

TODAY'S PROBLEM

Recently, a number of instances of fraudulent research reports, typically from premier institutions in the United States, received wide publication in the popular press. Indeed, in just the last few years, there has been an explosion of widely reported cases of scientific misconduct in the press.

In August 2010, as reported on the front page of the *New York Times,* Harvard University found Marc Hauser, a prominent researcher in animal cognition and morality, "solely responsible" for eight instances of scientific misconduct. Dr. Hauser's difficulties began when a research assistant accused him of subjectively influencing assessments of primate behavior. Based on this, in 2007 university officials went into his lab one afternoon while he was out of the country and publicly confiscated his records. Some of his data were found to be missing, and critics have since challenged Hauser's published results as incorrect or unconvincing.

In 2008, Linda R. Buck, a 2004 Nobel laureate in physiology or medicine for deciphering the workings of the sense of smell, retracted a 2001 paper that had been published in *Nature.* In September 2010, she retracted two more published papers, one from *Proceedings of the National Academy of Sciences* in 2005, one from *Science* in 2006. The retractions did not concern the work for which Dr. Buck won the Nobel Prize. However, the first author of all three retracted papers was a postdoctoral researcher who had conducted the experiments in Dr. Buck's lab. Dr. Buck was unable to reproduce the key findings in these papers; her figures were inconsistent with the original data.

In October 2010, three Harvard researchers retracted a paper published in *Nature* that had the far-reaching claim that the aging of stem cells might be reversible. Called into question was the reliability of the data from one of the authors, a postdoctoral student who performed the experiments in young mice and analyzed the results.

In 1998, Dr. Andrew Wakefield and his colleagues published an article in *The Lancet* that linked childhood vaccinations to autism. In January 2011, the study was renounced by 10 of the 13 authors and retracted by the medical journal. An editorial in the medical journal *British Medical Journal* has denounced the study as "an elaborate fraud." While the other scandals I've mentioned involved perhaps more esoteric subjects, this particular piece of fallacious research affected millions of parents in the United States who, fearing the contagion of autism more than that of chicken pox, diphtheria, whooping cough, and measles, prohibited their pediatricians from inoculating their young children. One can only speculate on whether the fears stirred up by this discredited bit of science have led to the current mini-epidemic of measles. (According to the CDC, more cases of measles were reported in 2008 than had been in any other year since 1997. Over 90 percent of those infected had not been vaccinated or their treatment status was unknown.) And what of the angst and self-blame that parents of autistic children must have felt?

The vaccine/autism affair makes clear that these kinds of scandals don't merely exist in the ivory towers of academe or in the white-coated confines of the lab. They have real-life repercussions for all of us.

In these high-profile retractions, it is unclear whether the highly charged, competitive nature of science puts pressure on researchers, causing everyone from the postdoctoral student to the primary investigator to cut corners, or whether the laboratory or clinical chief creates an atmosphere that induces cheating. But whatever the cause, now there is no question that the impassioned race in scientific research for recognition and rewards—with its bitter conflicts over priority and credit—has been brought to public awareness.

THE TRUTH ABOUT SCIENTISTS

The scientist is generally viewed as detached, objective, dispassionate. Nothing could be further from the truth. The coin of the realm is originality and creativity, but the joy of discovery, the exaltation of uncovering a law of nature, while a powerful motivation, is not enough.

Science grows out of an intense compulsion to understand, to make sense of the world around us. Carl Sagan eloquently framed the issue: "For myself, I like a universe that includes much that is unknown and, at the same time, much that is knowable. A universe in which everything is known would be static and dull...A universe that is unknowable is not a fit place for a thinking being. The ideal universe for us is one very much like the universe we inhabit. And I would guess that is not really much of a coincidence."[3]

Yet while discovery provides extreme excitement, scientists also covet recognition for their work. The reward may be not only high regard by peers, but a prize, an acknowledged victory. This is an intrinsic part of the field's culture, but it's not one scientists advertise to the general public.

The scientific enterprise brims over with competition, battles, and injustices. Conflicts may be resolved in an amicable fashion or may ignite bitter recriminations. Scientists are as subject to pride, greed, jealousy, and ambition, just like the rest of us. The race to be first can lead to superhuman effort and can, perhaps, even speed up the desired result. Disputes over who got there first often reveal how temperament, personal ambition, and antagonisms all too often influence the course of events.

AHEAD OF ITS TIME

A lot of the problem starts with the fact that the public is not always prepared to accept a major scientific discovery. Scientists themselves may forcefully deny the value of a new concept that clashes with accepted facts. Such resistance and neglect contradict the stereotype of the scientist as an open-minded searcher of the truth. An accomplishment labeled "ahead of its time" may be rejected simply because it was produced before it could be generally understood or before it could be technically confirmed. This can create open conflict when the scientist who discovered the new concept supports his discovery with unshakeable certainty.

In 1632, Galileo defended the heliocentric Copernican system in disregard of the Church's admonition. He was finally tried by the Inquisition and under threat of torture recanted. Legend has it that as he left the tribunal, he murmured under his breath regarding our planet, "Eppur si muove!" ("And yet it moves!"). Old conflicts die hard: The sentence passed on Galileo by the Inquisition was formally retracted by Pope John Paul II only on October 31, 1992.

The work of Gregor Mendel laid the foundation of modern genetics, but its importance went unrecognized for over forty years.[4] Mendel's prophetic remark, "My time will come," reveals no doubt on his part.[5] In the history of modern biology, Mendel's article is probably secondary in importance only to Darwin's "On the Origin of Species."

Barbara McClintock finally won the Nobel Prize for medicine in 1983, thirty years after she began presenting her revolutionary work on genes jumping within a chromosome. Scientists simply could not reconcile her conclusions on transposable elements with the prevailing belief in the stability of genes on the chromosomes. As one expert explained, "it was so far out that no one could relate it to anything anyone at the time knew about the genome."

Fifty-five years after his breakthrough discovery of a cancer-inducing virus, Peyton Rous, at age eighty-five, was awarded the Nobel Prize. In response to his 1911 paper in the *Journal of Experimental Medicine*, the official publication of the Rockefeller Institute, critics scoffed that there must have been a hole either in the filter he used for his supposed cell-free experiments or in his head.[6] His insight had to await a favorable intellectual climate and technologic advances.

Central to the narrative I present in *Prize Fight*, especially in Part II, are two emblematic breakthroughs that earned each major researcher a Nobel Prize, yet resulted in acrimonious conflicts over priority and credit. In one case, the "hurt" and efforts at rehabilitation lasted fifty years and in the second, thirty years. Each story is marked by raw competitiveness and guile, with dramatic turning points. Both conflicts were considerably unprecedented in their own eras due to the very public nature of the disputes.

The first story involves Selman Waksman, an internationally renowned scientist at Rutgers University. Waksman began a screening program for antibiotics derived from soil microbes, literally from the ground he walked on. A young graduate student in Waksman's laboratory, Albert Schatz, uncovered a new antibiotic, streptomycin, which was proved by clinical trials at the Mayo Clinic to be the first antibiotic effective against tuberculosis. The discovery was hailed as a major life-saving advance. This wonder drug, along with penicillin, opened the antibiotic era. Schatz was the senior author of the reports and shared the patent with Waksman, who nevertheless undertook a series of stratagems to marginalize Schatz. In an increasingly bitter exchange of letters, Waksman viewed Schatz as "my hands, my tools" as well as "a cog in a wheel" of his own fundamental research. Schatz tried to counter this, as well as Waksman's threats to his career, with his own ruses. The most important thing to him was ownership of his work. He wanted recognition for his research.

The second narrative involves my colleague, Paul Lauterbur, a physical chemist at the State University of New York at Stony Brook. Lauterbur had an epiphany, a "flash of genius," that transformed a basic technique used by chemists and physicists, nuclear magnetic resonance (NMR), into a clinical diagnostic method now employed worldwide—magnetic resonance imaging (MRI). Today it is a billion-dollar industry with more than 60 million medical diagnostic scans performed worldwide each year. Who has not had or known someone who has had an MRI study? Lauterbur was awarded the Nobel Prize in 2003.

These facts alone do not suggest the underlying fiery dispute that raged for thirty years around the development of MRI. Raymond Damadian, a physician-researcher at the SUNY Downstate Medical Center in Brooklyn, had published the seminal paper that first directed the attention of chemists to the vast medical potentials of NMR. Lauterbur cited this report in his grant applications for funding but chose not to cite it in his first paper introducing the procedure's imaging capability to the scientific community. Damadian became convinced that Lauterbur was exploiting his work.

My sources include extensive archival material from Rutgers University in New Brunswick, New Jersey, and Temple University in Philadelphia, containing voluminous correspondence, memos, personal reflections, pre-trial depositions, photos, and movies; and contemporary accounts, memoirs, extensive in-depth interviews, as well as my direct personal involvement in Lauterbur's productive years.

Basic to the integrity of scientific research are issues of intellectual property rights, peer review, and authorship related to allocation of credit. As an author of over 200 scientific papers and an editor of medical books and journals, I deal with these issues of peer review, authorship, and credit every day. These are important not only for the approbation by peers but also for much more practical reasons: They lead to promotions, funding, tenure, and status. Indeed, I had become particularly sensitized to the highly competitive nature of scientific investigation because of my direct encounter with the Lauterbur/Damadian conflict at the medical school where I taught. This led me on a quest to understand such behavior among scientists, and in so doing I uncovered a little noted but fundamental pattern among my colleagues involving self-interest, competiveness, and the battle for recognition and reward. To be first is paramount. To be second is to be forgotten.

Complementing these two detailed cases are other profiles of discoverers that illustrate these dynamic human characteristics. They are most fully appreciated in the context of the underlying cultural as well as scientific factors. Among the subjects explored:

- Priority in the history of science. Who was the first to uncover or to interpret the meaningfulness of a new scientific truth? How conflicts underscore that proper credit is a bid for immortality, overcoming a fear of being written out of history.

- The creative impulse and similarities between science and art. Both are passionate enterprises that are generated by similar personal characteristics. How passionate an enterprise science is and how like artists scientists are.

- Changes in the depiction of scientists in popular books from *Arrowsmith* by Sinclair Lewis to *Intuition* by Allegra Goodman. How the perspective of the nobility of research has become modified over time by the goal of personal fame and glory.

- Flaws in the system of peer review that may stifle innovation with common rejection of what ultimately prove to be major advances.

- Deserving contributions that have gone unrewarded by awards committees.

- Examples of the dark side of science, fraud, which not only exploits the flaws in the bedrock of the discipline but also provides insights into the pressures of research coupled with the vanity of celebrity.

Science arises from the relentless urge to know. While competition may be beneficial to spur scientists toward discovery, it can also stymie progress. The inherent competition in scientific research, pervasive yet not frequently acknowledged, may stifle young minds and make them afraid to be innovative. This is because the peer review process is arbitrated by senior experts in the field who tend to follow the tail of the comet and may lack the imagination to recognize ingenuity. To understand the interplay of these factors within science benefits us greatly. It allows us to be wiser to the complexities— intellectual, technical, and human—in the scientific endeavor, the proper recognition of credit for achievements, the mentorship of new scientists, and the allocation of research funding and resources.

PART I

CHAPTER 1

STOLEN CREDIT: A UNIVERSAL OUTRAGE

There are some things which cannot be learned quickly and time, which is all we have, must be paid heavily for their acquiring. They are the very simplest things; and because it takes a man's life to know them, the little new that each man gets from life is very costly and the only heritage he has to leave.
— *Ernest Hemingway*, Death in the Afternoon

In the early part of the twentieth century, a bitter conflict erupted over credit for a momentous scientific discovery, that of insulin to treat diabetes. This conflict heralds many of the issues that underlie disputes to this day.

Disordered breakdown of the body's circulating sugar creates the manifold and often lethal complications of what is known as diabetes mellitus. By 1920, the idea was growing that certain groups of cells in the pancreas, known as the islets of Langerhans, might be important to blood sugar metabolism and function independently from the rest of the pancreas cells.

A novice in the field—an inexperienced researcher who was unaware of virtually all of the previous experimental work—attempted to prove this theory with a novel approach. As Michael Bliss notes in his incisive account of this case, Frederick Banting, a twenty-eight-year-old country doctor, set out to isolate undamaged islets of Langerhans in order to extract their internal secretions, that is, a hormone.[1] He did this by tying off the gland's main duct in order to destroy the preponderant pancreatic cells producing the digestive ferments.

Banting approached John Macleod, chairman of the physiology department at the University of Toronto where Banting had obtained his medical training. A more mismatched couple could not be imagined. Macleod, an urbane and distinguished academician, doubted that the cause of diabetes mellitus was hormonal. But he recognized that the experiment that the raw, unsophisticated, and inarticulate—but intensely determined—young doctor proposed was a sound one, even if it were to provide a negative result. So he finally offered Banting laboratory space, some experimental dogs, and the assistance of Charles Best, a young graduate student.

In mid-May 1921, the two started their experimental work on dogs. Conditions were sweltering. They worked in a small, dirty operating room just next to the odoriferous animal quarters, under the gravel-and-tar roof of the University of Toronto's old medical building. In mid-June, Macleod left on summer holiday in Scotland for three months. Initially, many of the dogs died, often from shock and widespread infection. Banting's research program was facing total failure. To compensate for the loss of the university's dogs, he bought others on the streets of Toronto, occasionally leading them back to the lab by his tie. By the end of the summer, Banting had successfully isolated a pancreatic filtrate that substantially improved blood glucose levels in diabetic dogs.

Toward the end of September, Banting demanded more resources from Macleod—a salary, better facilities, an animal caretaker—under the threat of going elsewhere. Macleod was reluctant but acquiesced. This episode engendered ill feelings that heralded Banting's deepening hostility and insecurity.

At the annual meeting of the American Physiological Society at Yale University on December 30, 1921, Banting presented the group's results. He was nervous and spoke haltingly, admitting later that he was "almost paralyzed" and "overawed" before the dignitaries in the audience. Macleod joined the discussion, responding to serious criticism of the research, speaking smoothly and dramatically, and referring to "our work." This only added to Banting's suspicion that Macleod was trying to take credit. Macleod, he was convinced, "had no original ideas. He only knew what he read or was told and then he could rewrite or retell it as though he were a scientist and a discoverer."

Banting and Best wrote up the results of their work, which were published in a paper boldly entitled, "The Internal Secretion of the Pancreas," in the *Journal of Laboratory and Clinical Medicine* in February 1922. The article concluded, "we feel justified in stating that this extract contains the internal secretion of the pancreas."

The pressing goal now, as it always had been in Banting's mind, was to isolate and purify the extract for clinical use in the treatment of diabetes. To this end, Macleod assigned to the laboratory the service of an expert biochemist, James Collip, who was a year younger than Banting. Within weeks, Collip developed a way to purify what was to be known as "insulin," a name based on the Latin root for the word "islands." "I experienced then and there...perhaps the greatest thrill which has ever been given to me to realize," Collip exulted.

In January 1922, a fourteen-year-old boy, dying from diabetes at Toronto General Hospital, was given insulin. His rapid improvement was astonishing. This dramatic interruption of a fatal metabolic disorder would go down in the annals of the history of medical science as a medical miracle and would be forever associated with the names of Banting and Best.

At this point, however, a schism in the team effort created a personal confrontation—also remarkable in the history of science. Collip declared his intention to patent in his own name the method for purifying the extract. Banting, who regarded Collip as one of Macleod's allies, was so enraged, in one account, that he "suddenly and unexpectedly knocked [Collip] cold." This episode not only became scuttlebutt among contemporaries, but it has also entered medical legend. On a positive note, however, it promptly led to an agreement whereby the U.S. patent was assigned to the University of Toronto, which could license firms for the manufacture and sale of insulin and collect royalties. A similar arrangement was made with the British Medical Research Council.

On May 3, 1923—only fifty weeks after Banting and Best started their work—Macleod presented a summary paper before the Association of American Physicians in Washington, D.C. This served to widely announce the discovery of insulin. The breakthrough was hailed as "epoch-making" and "one of the greatest achievements of modern medicine."

In an effort to record what each had contributed, and as a further indication of bitter divisiveness, the Banting/Best team and Collip published separate reports describing the methods for making the insulin-containing extracts. Everyone, including Macleod, had developed grudges, suspicions, and misunderstandings regarding appropriate credit for the discovery of insulin, leading to an irreconcilable rupture.

By mid-September 1923, the clinical program was a huge success with strikingly dramatic results. Up to 25,000 diabetics were receiving insulin, and terms like "resurrection" and "salvation" were being applied to the seemingly miraculous recovery of debilitated and comatose patients.

On October 25, 1923, the Nobel Assembly—in an unusually rapid decision—awarded the Nobel Prize in physiology or medicine to Banting and Macleod. When Banting heard that Macleod was a co-recipient and Best had not been included, he was furious. He had always considered Best not as an assistant but as a partner in his work. He immediately decided to share his award with Best. Macleod, learning of this, announced his intention to share his prize money with Collip.

The acrimony Banting felt toward Macleod flowed from Banting's deep sense of injustice. He had undertaken his experiments with a fresh, albeit untutored, mind and with imagination, unshakeable faith, and determination. Macleod, on the other hand, felt that he had offered encouragement, advice, and resources. Furthermore, having been trained in Europe, Macleod probably had incorporated the hierarchical attitude that work successfully accomplished in his department should bear his co-authorship.

A fundamental issue throughout this episode is the issue of intellectual property. Who has priority? The discoverer, or the individual who helps to illuminate the concept to others?

THE ECSTASY OF DISCOVERY,
THE AGONY OF STOLEN CREDIT

Many of us have had a fresh idea or proud accomplishment appropriated by somebody else—perhaps a boss, a manager, a teacher, or a colleague.

An original idea, a novel approach, a creative effort—whether in the arts, the sciences, technology, or business—stirs not only pride but also an intense proprietary attachment within an individual. The passions aroused by claims over credit are acute. Credit is, after all, the basis for getting recognition, jobs, promotions, and awards. Its appropriation by another, particularly without attribution, is an act of plagiarism, a violation of intellectual property. The hurt this engenders in the originator results in a deep and long-lasting scar.

Although the practice of supervisors claiming credit for the work of subordinates is widespread, most complaints are voiced privately. Grumbling is generally restricted to the locker room or the water cooler.

Going public with such complaints is far less common. The accuser may not voice his or her distress for a long time and may express righteous indignation or seek passionate retribution.

Consider these examples:

The Polio Vaccine

On April 12, 1955, following nationwide clinical trials on 1.8 million American school children, the Salk vaccine against polio—at the time the leading crippler of children—was announced as "safe, effective, and potent." Euphoria swept the country. As Richard Carter notes in his biography of Jonas Salk, "people observed moments of silence, rang bells, honked horns, blew factory whistles, fired salutes, kept their traffic lights red in brief periods of tribute, took the rest of the day off, closed their schools or convoked fervid assemblies, therein drank toasts, hugged children, attended church, smiled at strangers, forgave enemies."[2]

In his role as director of the laboratory, Salk had been coldly distant from his staff, communicating through notes and memos, but was adept at self-promotion in his dealings with reporters. But an incident in 1954 showed that he was capable of duplicity. Julius Youngner, Salk's top research assistant, developed a critical test to measure the amount of polio virus in living tissue culture and handed a draft of the report to Salk for his comments. The original report listed only Youngner and an assistant as authors. Youngner asserts that Salk had done nothing to initiate, advise, or carry out the work. After a few days, Salk returned a "revised" draft to Youngner, but with Salk's name as first author. He claimed that he had lost the original draft, explaining that he nevertheless found Youngner's tables of data, from which he had to reconstruct the entire paper. On this basis, he made it clear that he considered it only fair that his name go first.[3]

At the formal announcement of the vaccine's success the following year, attended by more than 150 reporters, Salk's coworkers sat proudly together, expecting to be honored. They were, however, painfully snubbed and felt betrayed.

More than thirty-eight years later, Julius Youngner, then a distinguished service professor in the University of Pittsburgh's School of Medicine, had an opportunity to release the hurt he had long been harboring to Jonas Salk. He said to his former boss: "Do you still have the speech you gave...in 1955? Have you ever reread it? We were in the audience, your closest colleagues and devoted associates, who worked hard and faithfully for the same goal that you desired...Do you realize how devastated we were at that moment and ever afterward when you persisted in making your co-workers invisible?" He then asked Salk if he understood what he was saying, and Salk answered that he did. It's not clear if the response gave Youngner any solace.[4]

A Breakthrough Antidepressant

"Scientists, like everyone else," an editorial in the *New York Times* commented, "are motivated by more than the search for truth. They lust after reward and recognition and frequently dispute the credits that are distributed."[5] The editorial was reflecting on the outcome of a long and bitter conflict surrounding iproniazid, a breakthrough antidepressant of the 1950s.

In the early 1950s, Nathan Kline, a bold and energetic psychiatric researcher and director of the massive Rockland State Hospital in Orangeburg, New York, had introduced the use of tranquilizers to the practice of psychiatry, which steered the specialty onto a whole new path that focused on the chemistry of neurotransmitters in the brain. For this he received the prestigious Albert Lasker Medical Research Award in 1957. He went on to oversee clinical trials with the antidepressant iproniazid on over 800 patients at Rockland, and a paper attesting to its positive effects was read by Harry Loomer, the physician in charge of the Rockland clinical unit, at the regional meeting of the American Psychiatric Association in Syracuse, New York, in

1957.[6] However, the *New York Times* incorrectly credited the presentation to Kline. This probably happened because at a press conference in New York a few days earlier, details of the success of the trial had been provided solely by Kline.

When a scientist presents the common work of a group in which he is a co-worker, but mentions only his own name and none of the co-workers, it crosses the line from a simple violation of etiquette to the perception of scientific dishonesty. Loomer was infuriated, and the *New York Times* printed a retraction. Another member of the team, John Saunders, a physician and clinical pharmacologist who supervised the project and had correctly attributed the mode of action of iproniazid to monoamine oxidase (MAO) activity, also felt exploited. The following year, under his single authorship, Kline published a paper on the clinical experience with iproniazid.[7] Within one year of Kline's report, more than 400,000 patients had received the drug for treatment of depression. The spark of discontent burst into flame in 1964 when Kline became the single recipient of a second Lasker Award for work with iproniazid. (The accompanying citation said: "Literally hundreds of thousands of people are leading productive, normal lives who—but for Dr. Kline's work—would be leading lives of fruitless despair and frustration.")

Loomer and Saunders objected, saying they had made the discovery, and brought suit, asking for $1 million in damages from Kline for "false and fraudulent" claims for complete credit and from the Lasker Foundation for failure to make a "proper investigation into the facts." Seventeen years of litigation followed. Their decision to undertake the suit is all the more startling in the face of Kline's accomplishments and reputation. At some point, Loomer dropped out. Iproniazid had been introduced in clinical use originally as an anti-tuberculosis drug, and Saunders maintained that it was he who observed its euphoric effects and realized its potential. A jury ruled for Saunders, and the State Court of Appeals upheld that ruling and finally laid the case to rest in 1981. Saunders received one-third of the $10,000 Lasker award from 1964, and Kline was required to pay certain court fees totaling about $20,000.

This research culminated in today's multi-billion-dollar mood-altering drug industry. These MAO inhibitors have been replaced by the now-dominant class of antidepressant compounds, the tricyclics.

A New Radionuclide

A precedent-setting case that shook the halls of academic medicine in the late 1980s through the early 1990s established the right that faculty members have to their intellectual property. The case flared with issues of plagiarism, sex discrimination, due process, and institutional ruthlessness.

Something of a wunderkind, Heidi Weissmann started college when she was fifteen, began medical school when she was nineteen, and by age thirty-three, her supervisor was calling her one of the nation's premier authorities on biliary imaging in nuclear medicine. As an assistant and subsequently associate professor at the Albert Einstein College of Medicine of Yeshiva University, and an attending physician at its teaching hospital, the Montefiore Medical Center in Bronx, New York, she pioneered the clinical usefulness of a new radionuclide, iminodiacetic acid (IDA), for imaging the gallbladder and biliary system. This quickly proved to be a major advance in the diagnosis of cholecystitis (inflamed gallbladder), gallstones, and tumors of the bile ducts.

Weissmann published a series of articles detailing its value, often in collaboration with her mentor at Montefiore, Leonard Freeman, a highly respected professor of radiology and chief of nuclear medicine. Joint "honorary" authorship is frequently demanded in the sciences by the director of a department or laboratory on the basis that they have provided space, resources, and funding. According to Weissmann, Freeman made it clear that "he expected his name to be added" to any articles written by physicians working under his supervision, but it was Freeman who had contracted with the manufacturer of the isotope for funding the clinical research and offered Weissman the opportunity to pursue the project.[8]

Freeman thought so highly of Weissmann that in 1984 he recommended her for an award by the Society of Nuclear Medicine, referring to "the phenomenal amount of clinical research she has produced," adding, "I doubt that many other 30-year-old investigators could match her first-class productivity."

But over the next three years the relationship deteriorated. In January 1987, while Weissmann was on sabbatical, she received a letter from the department chairman that her return in September would be on a "probationary basis." Weissmann found this unsettling, given that she had been with the department for a decade. Lacking board certification in the field despite a career of several years, she was passed over for advancement in favor of a male faculty member. In April, she filed a complaint with the Equal Employment Opportunity Commission, alleging sex discrimination. Relations between her and her seniors were headed for the breaking point. In August, as the end of her sabbatical drew near, Weissmann was shocked to discover that Freeman planned to present a chapter she had published in a course he was teaching at another medical center in New York City, and that he had simply substituted his own name for hers on the work.

"I had accepted adding his name gratuitously to other publications where I was the primary researcher," she says, "but here he simply had his secretary white-out my name and reproduce the full text for distribution at the course under his name alone. My accomplishment was being obliterated."[9] Freeman reasoned that this involved "only a handout" included in a syllabus for distribution at a course.[10] Through a threatening letter from her lawyer, Weissmann demanded that the material be withdrawn from the course materials, and she filed suit against Freeman for copyright infringement. Since plagiarism is not a legal charge, monetary damages were not sought. An article in the *Chicago Tribune* later applauded the attack against the "medieval" guild structure of academia.[11]

Later that month, on the day that Weissmann and Freeman were scheduled to go to court, Weissmann was locked out of her office and escorted off the medical center's premises by two security guards. Two months later, she was fired. Freeman was promoted to vice chairman of the nuclear medicine department.

In March 1988, the case was heard as an issue of intellectual property rights by the District Court for the Southern District of New York, without a jury, before Judge Milton Pollack.[12] He was an eighty-two-year-old judge with a reputation for taking sides early, giving no quarter to the other side, and pushing the case to his chosen result.[13] He apparently cast certain academic faculty members in a very narrow role: "Just as he had his law clerk actually write his decisions," Weissmann asserts, "so he viewed an assistant professor in a faculty setting as merely a clerk to the professor." Nor was Weissmann assuaged by a headline in the *New York Times*, referring to her as an "aide" to a "top doctor."[14] Nevertheless, the article did refer to "plagiarism" as the accusation. Pollack asserted that Freeman's role was to lend "credibility" by his "reputation, knowledge, perception and experience." Pollack credited Freeman as the principal investigator who made the research possible; Freeman's stance was that he had shared authorship with Weissmann on several occasions and the revisions and updating were "trivial."[15] The case, Pollack concluded, was a matter of "misguided ego" on Weissmann's part.

About a year later, the Second Circuit Court of Appeals came to what some may consider a more enlightened decision. Judge Richard Cardamone observed that Weissmann's modifications must have been nontrivial; otherwise Freeman would not have copied the review but used an earlier version of the work. He pointedly noted that "in an academic setting…what is valuable is recognition," which, as the basis of incentives of promotion and advancement, is the "fruit of one's labor" and acknowledged that Weissmann was the proprietary author of the contested work. Freeman's reputation and authority were rejected as sources of authorship. The case was resolved in Weissmann's favor. The decision pointed out that "Dr. Freeman stood to gain recognition among his peers in the profession and authorship credit with his attempted use of Weissmann's article; he did so without paying the usual price that accompanies scientific research and writing, that is to say, by the sweat of his brow."[16] It was made clear that all that a scholar has is his or her reputation, based upon appropriate credit.

Though Freeman was found guilty only of copyright infringement in a case that many academic clinicians felt was frivolous and overblown, some people contend that his actions clearly constitute plagiarism. The case supported the importance of copyright protection to scientific production.

Underlying this conflict was a sex discrimination complaint filed by Weissmann against Montefiore around the same time as the copyright infringement suit against Freeman, alleging that she had been denied equal pay, promotions, and raises because she was a woman.

In September 1989, the Albert Einstein College of Medicine, Montefiore Medical Center, Yeshiva University, and Dr. Freeman proposed a settlement of $150,000 for the concurrent sex discrimination lawsuit if Weissmann dropped her charges of copyright infringement and advised the Congressional committee investigating academic misconduct that Freeman "did not engage in any scientific misconduct or breach of ethics." Weissmann refused the settlement offer. Freeman petitioned the U.S. Supreme Court to review the appellate court's findings, but was denied.

The lawsuits set off reverberations among other scientists, major philanthropic donors to Yeshiva University, and Congressional investigators dealing with alleged misconduct by scientists.[17] The Feminist Majority Foundation assisted Weissmann in the lawsuit. The National Coalition for Universities in the Public Interest, an organization founded by Ralph Nader, placed an ad on the op-ed page of the *New York Times* in May 1990, asking, "Why is the victim being punished and the perpetrator rewarded?"[18] In its September 1990 report, the subcommittee of the Human Resources and Intergovernmental Relations, chaired by Rep. Ted Weiss, found that both Montefiore and Einstein deliberately took the side of Freeman and ignored their own rules about academic misconduct, going so far as to disband a faculty senate committee appointed to investigate the issue.[19]

Throughout this time, Weissmann suffered the scarlet letter of a whistle-blower. Like many before and since, Weissmann paid a price for it. She was unable to find professional work in academia or private practice.

Finally, in 1994, after seven years of legal battles, Montefiore and Einstein settled out of court for $900,000 for sex discrimination and wrongful discharge.[20] Weissmann had spent $575,000 in legal fees. She used part of the award to establish a network and resources center to help other women and whistle-blowers who face similar problems.

Today, Heidi Weissmann still shakes her head in disbelief: "I thought one would be dealt with fairly in academia. I was so naive."[21]

EYES ON THE PRIZE

It is through originality that knowledge advances. On this basis, recognition and esteem accrue. Robert Merton, the eminent sociologist of science, has emphasized that the joy of the discovery alone is not enough.[22] A discoverer—no matter how he comes upon the finding—can say to himself, "This is mine. This is my grain of gold." It is not simply egotism, but rather reflects the scientist's inchoate longing to be recognized by his peers. As Darwin once phrased it, "My love of natural science... has been much aided by the ambition to be esteemed by my fellow naturalists." This motivation does not change with time. Citations in the literature of seminal or notable papers continue generally for about nine years and then begin to trail off as the contributions become incorporated in the body of knowledge. They become "household words" no longer requiring attribution.

The Nobel Prize is generally acknowledged as the top award of recognized achievement in science. Beyond this is an array of medals, memberships in honorary academies, fellowships in prestigious societies, and invitations to present highly regarded lectures. Historians of science are also important in maintaining recognition, both during an individual's lifetime and posthumously.

Alexander Fleming reportedly confessed to a scientific contemporary in 1945 that he didn't deserve the Nobel Prize but added that he nevertheless couldn't help enjoying his fame. He spent the next ten years of his life collecting twenty-five honorary degrees, twenty-six medals, eighteen prizes, thirteen decorations, and honorary membership in eighty-nine scientific academies and societies.

A Nobel committee representative presented a gift to Rosalyn Yalow, the scientist who developed a technique to identify traces of substances in the human body, of a small replica of the Nobel medal. She made a necklace of it and, on occasion, proudly wore it in her lab at the Bronx VA Hospital.[23]

Not all laureates bask in the glow of the award. For most, it is a mixed blessing, since it presages further changes in their work and their relations with other scientists and with the wider society. Ramón y Cajal, laureate in medicine for 1906, found the experience excruciating: "Months were squandered in acknowledging felicitations, in pressing friendly or indifferent hands, concocting commonplace toasts, recovering from attacks of indigestion and making grimaces of simulated satisfaction."[24]

The blunt form letter devised by Francis Crick, one of the two renowned scientists credited with recognizing the structure of DNA, leaves no doubt about its intention[25]:

> Dr. Crick thanks you for your letter but regrets that he is unable to accept your kind invitation to:

send an autograph	*help you in your project*
provide a photograph	*read your manuscript*
cure your disease	*deliver a lecture*
be interviewed	*attend a conference*
talk on the radio	*act as chairman*
appear on TV	*become an editor*
speak after dinner	*write a book*
give a testimonial	*accept an honorary degree.*

Many laureates wish simply to continue their lives. Tsung-Dao Lee, the scientist who explained one of the basic forces of the atom, was only thirty-one when he got the news of his Nobel Prize in physics in 1957. Rather than being elated, he needed cheering up. "My God," he said, "what happens now to the rest of my life?"[26] Werner Forssmann had to adjust to his new celebrity status when he won the 1956 Nobel Prize in medicine. Plucked out of obscurity a quarter century after his exploits, Forssmann, then a country doctor in the Black Forest, told a reporter, "I feel like a village parson who has just learned that he had been made bishop."

As for the Banting/Best, Macleod/Collip controversy, it should be noted that the wisdom of giving Macleod a share of the 1923 award as well as the question of Banting's scientific abilities were debated long after the prize was bestowed. Perhaps most righteously, by 1950, Best could derive personal satisfaction when the official history of the Nobel Prizes, in an unprecedented move, acknowledged that an error in judgment had been made in 1923 and that he ought to have shared the prize.

Finally, although the rewards of recognition are immensely varied, the use of eponymy—naming a discovery after its discoverer—heads the list.[27] In this way, scientists become immortal, with their names indelibly inscribed in all the scientific languages of the world. Examples abound, and some have so entered our everyday speech that little thought is given to their origin: Eponymy may range from the institution of an epoch (the Copernican revolution, the Freudian age); a new science or branch (Euclidean geometry); laws, instruments, and constants (Doppler effect, degrees Fahrenheit, Geiger counter); anatomic parts (Fallopian tube); diseases (Hodgkin's disease); diagnostic tests (Pap smear); and electrical and magnetic units (volts)—the latter having now become the name of a hybrid electric Chevrolet. And in a case of seemingly double eponymy, the Chevy itself was named after Louis Chevrolet, one of the original founders of General Motors. What's in a name, anyway?

CHAPTER 2

THE ART OF SCIENCE

Is there any similarity in the actual creative act between art and science? Yes, the
fact that in both you're doing what hasn't been done before. You flatter yourself
that you are the first to open up some terrain, or see something in a new way.
—*Carl Djerassi, biochemist, novelist, and playwright*

Our lives have benefited enormously from scientific and medical discoveries in the past century. The history of these advances is generally seen as a chronicle of sequential events without specific consideration of the underlying creative impulse or the human drama involved. Modern textbooks portray medical science as a set of current facts and strong beliefs, inscribed like the ten commandments by some distant authority. Space is rarely devoted to recounting how an inspiration occurred, nor to the often painstaking process involved. There is no grand theory of scientific discovery. Yet, once you penetrate the cool, dispassionate surface of reported results, considerable creative heat arises. Common themes run through many of the individual medical advances of the past century: the pathways and patterns of creative thought, the subtle variables that influence the discovery process, the ironies of circumstance, the element of surprise and wonder, the epiphany leading to a breakthrough in understanding, the urge to establish priority. These factors are as ubiquitous as they are hidden. Journeying among them lands you squarely in the underlying human drama.

Typically, an advance occurs over time through the contributions of many with the power of cumulative knowledge. Most scientific progress is the cumulative effect of an assortment of smaller discoveries. But sometimes an individual with a far-reaching imagination emerges who, with determination and perseverance against painful odds, can overturn long-held dogmatic beliefs and construct a new paradigm. The initial lack of general acceptance of new ideas within the scientific community is glossed over in time, and profound social and cultural changes often result. As a British physician and investigator wryly observed in the early 1900s:

> There are three stages in the history of every medical discovery. When it is first announced, people say that it is not true. Then, a little later, when its truth has been borne in on them, so that it can no longer be denied, they say it is not important. After that, if its importance becomes sufficiently obvious, they say that anyhow it is not new.[1]

The British Library has recently compiled a hundred of the twentieth century's major scientific and medical discoveries.[2] However, the element of human character in these epoch-making moments is not accorded its pivotal importance nor is the complex nature of the interpersonal relationships leading to the discovery revealed in the original scientific literature. Scientific articles generally present their observations, data, and conclusions in a dry passive voice with excruciating precision. As well, too much is at risk for scientists or physician-investigators early in their careers to admit to all the factors that led to their achievements. Only years later might they testify to the contributions of such mind-turning factors as timely development by others, fortuitous happenstances, or exceptions to a premise.

Creativity in discovery implies both an awe of the understanding of the universe yet to be achieved and an aesthetic appreciation of the beauty in unlocking its mysteries. This was poetically expressed by no less a luminary than Isaac Newton:

> I know not what I may appear to the world, but to myself I appear to have been only like a boy playing on the sea-shore, and diverting myself in now and then finding a smoother pebble or a prettier shell than ordinary whilst the great ocean of truth lay all undiscovered before me.[3]

Thoughtful analyses of the imperatives of a creative individual have been offered: The creative person is uneasy at the detection of inconsistency, gaps, or lack of harmony and seeks to achieve some sort of unity. In his book *The Dynamics of Creation*, the British psychiatrist Anthony Storr considers such "discontent of dissonance" the most precious human attribute.[4] The humanistic psychologist Rollo May underscores this perception: "We express our being by creating. Creativity is a necessary sequel to being."[5] Indeed, another compelling motive is the urge to consolidate a sense of identity. The composer Aaron Copland expresses this clearly in his Charles Eliot Norton lectures of 1951-2, Music and Imagination: "And why is the creative impulse never satisfied: why must one always begin anew?...each added work brings with it an element of self-discovery. I must create in order to know myself."[6]

SCIENCE AND ART: SHARED CREATIVE IMPULSES

In a highly influential 1959 lecture at Cambridge University in England titled "The Two Cultures," C. P. Snow lamented that a great "gulf of mutual incomprehension" separates the world of science and technology and the world of the arts—particularly literature.[7] He argued that practitioners in both areas should build bridges to further the progress of human knowledge and to benefit society. Snow's perspective continues to be provocative more than fifty years later. He observed that both cultures are distorted by their insularity and he protested against the tribal philistinism among the learned, both scientists and those educated in classics and humanities. An educated person, he maintained, should know both the second law of thermodynamics and the works of Shakespeare. Each year it becomes abundantly clear that some degree of scientific illiteracy generally must be overcome to permit intelligent citizens to engage in the wider discussion and decision making that determine which roads we take in research and in applications, and which ethical considerations should inform them.[8]

Two ardent spokesmen offer surprising contrarian but no less insightful appraisals. W. H. Auden admired scientists as "men of action...who transform the world." His self-image suffered by comparison: "When I find myself in the company of scientists, I feel like a shabby curate who has strayed by mistake into a drawing room full of dukes."[9] More soberly,

Lionel Trilling has lamented that physics is too intellectually demanding for all but the few, and to be excluded from the aesthetic rapture that it plainly brings to the best physicists is an affront or, as he calls it, "a wound to our intellectual self-esteem." Perhaps he had in mind the awe of revelation, expressed by the astrophysicist Subrahmanyan Chandrasekahr as "shuddering before the beautiful."[10]

Often, at least in the popular mind, the concept of creativity is automatically linked with the arts rather than with the sciences. Literature, plays, and movies depict the artist as obsessed, struggling with unrestrained needs for self-expression. The scientist, on the other hand, is generally viewed as methodical and disciplined, in pursuit of a predetermined goal to which a certain methodology is adapted and applied.

The intense and obsessed artist has imprinted himself in the mythology and legends of our culture. A story about Sir Walter Scott illustrates this. One day, when he was out hunting, a sentence he had been desperately trying to compose all morning suddenly leaped into his head. Before it could fade, he shot a crow, plucked off one of its feathers, sharpened the point, dipped it in the bird's own blood, and recorded the sentence.[11]

In the twentieth century, Henri Matisse, bedridden in his villa near Nice during his recovery from abdominal surgery, could not restrain himself from using a bamboo stick with chalk at its tip to draw on his bedroom wall. Among scientists, the creative urge is no less compelling. Paul Ehrlich obsessively worked his way through 606 arduous experiments before he came upon the "magic bullet" for treatment of syphilis. The nineteenth-century mathematician Karl Gauss exhibited a similar crazed intensity as he worked on a difficult problem. Advised by a servant that his wife, who was quite ill, was about to breathe her last, Gauss could only mutter, "Tell her to wait a moment till I'm through."[12]

Particularly illuminating is the experience of Jacques Monod, the 1965 Nobel laureate in medicine and physiology. During World War II, he used his research, absorbing as it was, as a cover for underground activities during the German occupation. Illegal printing presses, forbidden newspapers, and weapons were hidden in rooms behind the experimental apparatus. Then, he was forced to go underground full time, staying away from any part of Paris where he might be known. But he missed his research so much that, at great risk, he smuggled himself into the laboratory at the Pasteur Institute whenever time allowed to continue his experiment. He knew that he might not live long enough to finish the research, that even if he finished it he couldn't publish it, and that he would likely never receive prestige or recognition. When asked why he did it, Monod simply replied, after a thoughtful pause, "Because. That's what I wanted to be doing—that's what my life was about!"[13]

Creativity in science shares with the arts many of the same determinants.[14] Common to both are imagination and intuition; the search for self-expression, truth, and order; an aesthetic appreciation of the universe; a perception of reality; and a desire for others to share their view of the world. J. Michael Bishop, Nobel laureate and recently chancellor of the University of California, San Francisco, clearly sees the commonality: "In many ways, art and science are kindred souls. Both arise from the same transcendent human qualities: ambition, imagination, creativity, intellectual daring, and the urge to discover."[15] The novelist Vladimir Nabokov bridges the tension between the rational and the intuitive in his observation that "there is no science without fancy and no art without fact."[16] The use by poets and novelists of the sound, emotion, and rhythm of words, metaphors, and analogies is not an unbridgeable schism from similar uses by scientists.

Let us reflect on a rapturous observation by a highly perceptive scientific mind. In the summer of 1841, Michael Faraday, generally considered the greatest of all experimental physicists, was on holiday in Switzerland and came upon a wondrous waterfall. At the base of each of a series of cataracts, the water was shattered into foam and then tossed into "water-dust" in the air. He jotted down these brief notes in his personal journal, a habit he had cultivated to record detailed observations of his extensive experiments:

August 12th, 1841—To-day every fall was foaming from the abundance of water, and the current of wind brought down by it was in some places too strong to stand against. The sun shone brightly, and rainbows seen from various points were very beautiful. One at the bottom of a fine but furious fall was very pleasant,—there it remained motionless, whilst the gusts and clouds of spray swept furiously across its place and were dashed against the rock. It looked like a spirit strong in faith and steadfast in the midst of the storm of passions sweeping across it, and though it might fade and revive, still it held on to the rock as in hope and giving hope. And the very drops, which in the whirlwind of their fury seemed as if they would carry all away, were made to revive it and give it greater beauty.[17]

The major image is the rainbow, a form without substance, yet strong and enduring. His description of this simple metaphorical image displays a poetic way of seeing and mode of thought, since the "spirit" is formed among the chaos by the light and the perspective of the observer. It is this mind that discovered the breakthrough concept of the electromagnetic field, which provided the basis for the later theories of James Clerk Maxwell and Albert Einstein.[18]

"There was a moment," Richard Feynman noted in his exploration of quantum electrodynamics, "when I knew how nature worked. It had elegance and beauty." A determinedly irreverent personality, he could not refrain from adding, "The goddamn thing was gleaming."[19] The search for beauty is as important to the scientist as it is to the artist. Simplicity, elegance, harmony—these attributes delight the scientist's aesthetics of beauty.

Both science and art require the making of images in the mind. In this regard, Nobel laureate Salvador Luria saw little distinction: "The imaginative moment is as creatively central to science as to poetry or figure art. The mind acts upon the natural world in the creation of knowledge in the same way as it acts on the elements of human sensibility in bringing forth a poem or a painting or a symphony."[20] Imagination incorporates, even within its linguistic root, the concept of visual imagery. Indeed, such words and phrases as "insight" and "in the mind's eye" are derived from it. David Bohm expressed this duality well: "Physics is a form of insight and as such it's a form of art."[21]

To a scientist, the greatest reward is the thrill of discovery, the moment you alone know something new. "Lifting the corner of the veil," in Einstein's phrase, to reveal a universal truth brings a sense of great joy associated with heightened consciousness, intense excitement, exaltation. Such moments are often shot through with metaphors of religious experience and even sexual pleasure.

The joy of discovery tinged with a sense of prideful power is revealed by an observation about Linus Pauling:

> I will never forget the picture of Linus Pauling in a meeting at McGill describing the creative process. He rubbed his hands in pure sensual satisfaction and his baby-blue eyes positively glittered: "Just think," he said, "I know something that no one else in the whole world knows—and they won't know it until I tell them."[22]

Gerald Edelman, Nobel laureate in medicine, speaks of "almost a lustful feeling of excitement when a secret of nature is revealed."[23]

"The most beautiful thing we can experience is the mysterious," said Einstein, for "it is the source of all art and science."[24] For Einstein, the sense of mystery became another name for God. This view has been deeply held by many scientists. The underlying impulses that we now pursue separately as science, art, and religion undoubtedly arose from the common desire to make sense of the world.

A creative scientist believes that fundamental laws of nature, of the workings of the human body, are decipherable by human intelligence. One must view this capability with awe.

SCIENCE AND ART: CRITICAL DISTINCTIONS

Despite many similarities in impulses, critical factors persist that distinguish science as science and art as art. Foremost is the need in science for reproducibility, for verification and validation. Science in Gerald Edelman's elegant phrase is "imagination in search of verifiable truth."[25]

Intuitions—based on a sense of the reality behind the appearance—always appear at the fringe of consciousness, not at the focus. An individual must grasp them for valuable ideas in the eddies and backwashes rather than in the main current of thought. Intuition is a cognitive skill, a capability that can make judgments based on very little information. An understanding of the biological basis of intuition—one of the most important new fields in psychology—has been elaborated on by recent brain-imaging studies.

In young people who are in the early stages of acquiring a new cognitive skill, the right hemisphere of the brain is activated. But as efficient pattern-recognition synthesis is acquired with increasing age, activation shifts to the left hemisphere. Intuition, based upon long experience, results from the development of neural networks upon which efficient pattern recognition relies.[26] The experience may come from deep in what has been termed the "adaptive unconscious" and may be central to creative thinking in discerning analogies and anomalies, similarities among differences and differences among similarities.[27]

Before an intuitive scientific insight can be incorporated, it must be rationally validated. Artistic intuition, in contrast, does not require validation. An artist may be very firm in the expression of his own personal sensibility. Robert Rauschenberg, responding to an inquiry from an art gallery exhibiting his work, telegrammed back: "This is a portrait of Iris Clert if I say so."[28] To an artist, truth, in addition to beauty, may be in the eyes of the beholder. But in science, imagination needs to be focused through a distinct lens.

It is this demand for verification that creates scientific progress. And progress is itself a feature that distinguishes science from art. Scientific knowledge and understanding expand ever more. As they advance, the original scientific papers are typically left behind as they are absorbed into the body of knowledge—they are rarely reread, no matter how contributory or even revolutionary they may have been. Quite unlike great works of art, they are ephemeral.

Art offers an experience filtered through individual consciousness that not only allows but fosters repeated viewings or hearings. But while art gives us lasting legacies in the form of plays by Shakespeare and concertos by Bach, it cannot make progress in the way science does. A science museum may lead the viewer from a display of Leeuwenhoek's primitive monocular microscope to one of an electron microscope, or from an exhibit on phrenology to one on the chemistry of mood-altering neurotransmitters. Art museums, on the other hand, are founded on the principle that one artistic creation is not more "progressive" than another. A painting by Jackson Pollock is not an advance over Raphael, nor a Brancusi sculpture an advance over a Cycladian one. Developments in art in all its manifestations have led to a broader array of subject matter and forms but, it cannot be claimed, to advances in artistic expression.

CHAPTER 3

STAKING THE CLAIM

There is one character trait… which is an intrinsic part of a scientist's culture, and which the public image doesn't often include: his extreme egocentricity, expressed chiefly in his overmastering desire for recognition by his peers. No other recognition matters.

—*Carl Djerassi*, The Bourbaki Gambit

Scientists may strive to understand the secrets of how the world works, but—being human—they are not without a lust for recognition. Fundamental to this through the ages is the passion to establish priority.

Assessment of the competence, significance, and originality of scientific work submitted for publication was instituted by the *Philosophic Transactions* of the Royal Society of London as early as the late 1600s. Robert Hooke, a pioneering microscopist, served as the Royal Society's first curator of experimenters. The motto of the society, *Nullius in verba* ("On the word of no one"), expresses the society's insistence on verification by observation or experiment, rather than by the voice of authority or tradition.

In the seventeenth century it was the custom for an author to publish an anagram some time before publication of his actual work in order to verify priority. The anagram was, of course, unintelligible until the author was challenged and construed its meaning. In some cases the letters of a significant sentence were arranged in alphabetical order. An example is Robert Hooke's *ceiiinosssttv*, which, properly arranged, makes *tensio sic vis*: Hooke's law (establishing the relationship between the stress and strain in elastic bodies). Hooke established his law in the 1660s but did not publish the anagram until 1676, and he waited until 1678 to reveal its meaning.

So obsessed was Hooke with establishing his priority beyond any doubt that he scoured through the journal's logs in search of "omissions of things and names," drawing lines through empty spaces so that "there may be no new thing written therein." This assured him that nobody in the future could interpolate reports of later discoveries so as to usurp his authorship. Hooke's temperament was melancholy and distrustful, but his contemporary Isaac Newton displayed a vindictive and merciless nature. Newton's famous remark in a letter to Hooke in 1675, "If I have seen further it is by standing on the shoulders of giants,"[1] is generally taken as an acknowledgment of Newton's debt to his predecessors, but it is now recognized as a backhanded and cruel jab at the hunchbacked Hooke. Newton fought several battles with Hooke over priority in optics and celestial mechanics.

Newton became involved in a notably acrimonious controversy with Gottfried Leibniz over who invented calculus. In his October 1676 letter to Leibniz, Newton included his celebrated anagram:

6 acc d a e 13 eff 7i 319n 4o4qrr 4s8t12 ux

The anagram concealed Newton's formulation of the fundamental theorem of the calculus. Years later, in 1712, a special commission of the Royal Society decided upon Newton's priority and condemned Leibniz as a plagiarist. "We reckon Mr. Newton the first Inventor" was the unanimous verdict. Only later was it revealed that the draft of the commission's report was written by the society's president, Sir Isaac himself!

In the latter half of the nineteenth century, the rise and applications of science and technology were promptly embraced by society. In 1869, the journal *Nature* was founded in London with the goal of informing the public and scientists of advances "made throughout the world." Its current worldwide circulation is about 67,000. *Science* was founded in Washington, D.C., in 1880 to foster communications among American scientists, and soon thereafter it was acquired as the official organ of publication of the prestigious American Association for the Advancement of Science (AAAS). Today, the journal has a circulation of about 140,000. A succession of Nobel laureates has served as president of AAAS in recent decades. In the field of medicine, scientists established the *New England Journal of Medicine* and the *Journal of the American Medical Association* in the United States and the *Lancet* and the *British Medical Journal* in England. Under a "peer review" system, independent qualified experts judge submissions for publication in all these journals.

In the decades before World War II, some European journals offered a device called the *lettre de cachet* (sealed envelope). This was used by a researcher who wanted to keep his results secret from other scientists while maintaining a claim of priority if his competitors should publish first. An article based on successful preliminary experiments could be submitted so that the journal editor would date it upon receipt without, however, opening it. When a competitor had published, or was about to publish, the same material, the researcher would then ask that the manuscript go through the editorial process. Even though his article would likely appear later, the original submission date the *lettre de cachet* bore would demonstrate priority.

GENEROSITY DOES EXIST

Charles Darwin, not a professional scientist but a well-educated gentleman of independent means, was moved to complete his magnum opus on evolution, *The Origin of Species,* in 1859, after learning that Alfred Russell Wallace, fourteen years his junior, had independently conceived the same ideas. Darwin feared, quite rightly, that his theory of evolution would be fiercely opposed by the scientific and religious establishment and had long delayed publication, but he was plagued with uncontrollable ambivalence: "I rather hate the idea of writing for priority, yet I certainly should be vexed if any one were to publish my doctrines before me."[2]

And then, in June 1858, the world crashed around Darwin. He was shocked to receive material from Wallace that confirmed all his fears that someone else could legitimately claim first discovery: "I never saw a more striking coincidence; if Wallace had my MS sketch written out in 1842, he could not have made a better short abstract! Even his terms now stand as heads of my chapters...So all my originality, whatever it may amount to, will be smashed."[3] In a striking example of simultaneous discovery, Wallace had similarly formulated the theory of natural selection and even conceived of the phrase "survival of the fittest," as had Darwin. Both had been inspired by Thomas Malthus's *Principles of Population,* which concluded that human populations always grew faster than their ability to produce food; it struck both Darwin and Wallace that under the struggle for existence, favorable variations would tend to be preserved and unfavorable ones destroyed.

Darwin was torn between his strong sense of integrity and his struggle for originality and recognition. At first, magnanimously, he considered stepping aside. Then, he considered publishing a short version of his long-standing text, "a dozen pages or so," but concluded that "I cannot persuade myself that I can do so honourably" and yet exclaimed, "It seems hard on me that I should lose my priority of many years' standing."[4] Darwin's friends in the scientific community took matters in hand and arranged for both papers to be read at a session of the Linnean Society, for the stated purpose of "not solely considering the relative claims to priority...but the interests of science generally."

Rather than engage in a priority dispute, Wallace nobly acquiesced to the superior definitiveness of Darwin's contribution. He proved to be uncommonly generous, giving his own book on evolution the title *Darwinism.* Wallace became a lifelong friend of Darwin and served as a pallbearer at his funeral. He has come down through history as "being famous for having been forgotten."

Among those who could be in a position to claim priority for a prize-winning achievement, largess is a rare spirit. A few other notable exceptions exist in the history of science. One speaks volumes about the integrity of Willem Einthoven, a physiologist at the University of Leiden in the Netherlands. After six years of work, Einthoven successfully developed the string galvanometer, a huge construction that is the forerunner of the electrocardiogram (EKG) and that recorded the first human electrocardiogram on November 18, 1902. Einthoven was rather clumsy with his hands and relied very much on the collaboration of his chief technical assistant, K. F. L. van der Woerdt. When Einthoven received the Nobel Prize in physiology or medicine in 1924, he wished to share the $40,000 award with van der Woerdt but soon learned the man had died. He sought out van der Woerdt's two surviving sisters, who were living in genteel poverty in a kind of almshouse. He journeyed there by train and gave them half of the award money.

In 1978, Robert Furchgott, an academic researcher in pharmacology at the State University of New York (SUNY) Health Science Center in Brooklyn, stumbled upon "an accidental finding as a result of a technician's error [which] completely changed the course of research in my laboratory."[5] The unexpected observation reversed a methodology he had been using for decades in his research.[6] In quickly reporting the new fundamental discovery, Furchgott magnanimously included the name of the technician who had committed the oversight among the co-authors of the abstract.[7] Furchgott had discovered that the lining of blood vessels naturally produced a gas, nitric oxide, that constituted a signaling molecule between cells in the body. This revolutionized vascular biology. As research progressed by others, nitric oxide was shown to regulate an ever-growing list of biological processes, including blood pressure, blood clotting, bacterial endotoxic shock, the immune system, intestinal motility, behavior, and some aspects of learning. The general public knows one of the useful mechanisms of nitric oxide as the product Viagra. In 1998, the eighty-two-year-old Furchgott was awarded the Nobel Prize.

In 1974, Georges Köhler, a twenty-eight-year-old postdoctoral fellow, joined César Milstein's Medical Research Council (MRC) Laboratory of Molecular Biology in Cambridge, England. Milstein was an internationally acknowledged leader in the field of antibody research. Antibodies are proteins produced by the lymphocytes of the body's immune system in response to attacks by foreign bodies, called antigens, that range from infectious bacteria and viruses to common allergens. Milstein, a quiet and reserved man, was an imaginative researcher, receptive to new ideas. One night, while lying awake in bed, Köhler conceived of a way to immortalize the antibody-forming cell by creating a new cell that could reproduce indefinitely and yield unlimited quantities of a single pure antibody, a sort of precise guided missile. Milstein was initially skeptical but nevertheless encouraged Köhler, who went on to execute his idea by Christmas of that same year: He fused a lymphocyte (which was making a known antibody) with a cancerous myeloma plasma cell to produce a clone that turned out the same pure antibody as the parent lymphocyte but had the longevity of the cancer cell.[8] The results are called monoclonal antibodies—far more pure, uniform, and sensitive in probing for their specific targets than regular antibodies.[9]

Milstein shortly approached the National Research and Development Corporation, which had a monopoly over MRC inventions, to patent the advance, but it notoriously wrote of the invention: "It is certainly difficult for us to identify any immediate practical application which could be pursued as a commercial venture." This laboratory technique went on to spark the creation of an international billion-dollar biotechnology industry. The applications of monoclonal antibiotics are extremely varied as both diagnostic tools and treatment: in diagnosing and fighting viruses; in treating disorders of the immune system such as rheumatoid arthritis, juvenile diabetes, and multiple sclerosis; in aiding in the purification of interferon; in matching transplant tissues; and, when fused with radioactive chemicals or toxins, in destroying malignant tissue in cancer patients.

Although Georges Köhler was the senior author of the 1975 seminal report in *Nature*, awards began to be showered upon César Milstein. In 1980, Columbia University gave its $20,000 Horwitz Prize to Milstein alone for inventing the technique. The General Motors Cancer Research Foundation, too, awarded its $100,000 Sloan Prize to Milstein the following year for the technique "which Dr. César Milstein developed."

According to Nicholas Wade of the *New York Times,* Köhler, a shy and gentle man, did not complain of these oversights but insisted on his own role in the discovery, that of conceiving and executing the experiment. It occurred in his first year of independent research. "I believe I was the driving force in it, but it is also true that I would not have thought about this problem in any other laboratory than César Milstein's and I wouldn't have been encouraged to do the experiment by anyone else but César Milstein."[10] Milstein had experienced repression of the truth in his native Argentina and was sensitive to repression of others. In 1984, proper tribute was made when the Nobel Prize was awarded to both Milstein and Köhler.

THE RACE IS ON

The imperative to establish priority for a discovery was certainly a driving factor for medical student Jay McLean at Johns Hopkins, as early as 1916. As World War I raged, he was assigned a research product under the direction of William Howell, the world's expert on blood clots. McLean job was to find a natural body substance to clot blood. McLean, however, stumbled upon the exact opposite: a powerful anticoagulant (blood thinner). When his mentor was skeptical, McLean placed a beaker of blood before him and added the substance. The blood never clotted, but Howell remained unconvinced. McLean rushed the report of his discovery into print to stake his claim and to indicate that Howell had played a subordinate role. The substance shortly became known as heparin (from the Greek *hepar*, the liver, to indicate its abundant occurrence in the liver). Heparin has become a standard in treating a variety of venous and arterial clotting disorders.[11]

Nothing sharpens a scientist's drive for priority more than the realization that he is in a race with another researcher. As time becomes of the essence, the scientist's temperament and character become challenged, and unsavory aspects may emerge. Roger Guillemin, working at the Baylor University School of Medicine in Houston and the Salk Institute in La Jolla, and Andrew Schally, working at the Tulane University Medical School in New Orleans, engaged in a bitter twenty-year competition, all in pursuit of an objective deemed illusory by the best experts. Each was independently pursuing research into unknown hormones that might be secreted by a part of the brain, the hypothalamus that governs the pituitary gland. This is the control center of endocrine function in the brain. Virtually all the hormones with manifold functions in the human body are under the regulation of the pituitary gland. For one experiment, Guillemin collected half a million sheep hypothalami from slaughterhouses; Schally, with smaller resources, made do with 100,000 to 200,000 pig hypothalami at a time, obtained from meat packing houses.

Nicholas Wade's book *The Nobel Duel* reveals both as men of towering ego and petty resentment, both senselessly driven not only by a thirst for knowledge but also for personal aggrandizement.[12] Each was highly secretive and notably reluctant to acknowledge the other's efforts. In a rare exchange of letters, they were provocative and insulting. Each managed lab teams that often broke up over issues of ego, credit, or double dealing. In 1977, Guillemin and Schally shared the Nobel Prize for their work demonstrating the secretion of peptide hormones by the brain. They wouldn't talk to each other at the ceremony but were, for the first time, obliged to shake hands.

The highest levels of government were also necessary to resolve a frenetic race between two international rivals. It began in 1983 when Luc A. Montagnier at the Pasteur Institute in Paris reported a newly discovered virus isolated from a patient at risk for AIDS. Following collaboration with Robert Gallo's lab at the National Cancer Institute in Bethesda, Maryland,

Montagnier and Gallo simultaneously published an article in *Science* claiming evidence that AIDS was caused by a retrovirus. After the initial discovery, the story turns ugly as the two teams ended their close collaboration and a race developed with each trying to prove its own virus strain was indeed the cause of AIDS. This led to a long, often acrimonious controversy over credit for the discovery of the AIDS virus. Not only personal but also national issues of reputation and revenue were at stake. Gallo describes the dispute as "involving legal, moral, ethical, and societal questions that soon spilled over into the world of scientific research and threatened to poison relationships between scientists as well as between the research community and the general public."[13] Following a mishandled press conference, an editorial in the *New York Times* commented, "The commotion indicates a fierce—and premature—fight for credit between scientists and bureaucratic sponsors of research. Certainly," it concluded, "no one deserves the Nobel Peace Prize."[14] At one point Gallo was accused of fraudulently presenting Montagnier's virus samples as his own.[15]

In 1987, President Reagan and Prime Minister Jacques Chirac of France attempted to end the dispute and signed an agreement that each party had equal rights to claim priority concerning detection and isolation of the virus and to share patent rights to the technology for detecting infection with the virus; the tests are now used to screen blood donations, making the blood supply safer for transfusions and blood products. In 1991, Gallo acknowledged that the AIDS virus he had "discovered" in 1984 really came from contamination of his cultures by the virus from a sample from the Pasteur Institute. The two teams had regularly swapped samples. Gallo was cleared of any wrongdoing, and in 2010, Montagnier was awarded the Nobel Prize in medicine. Today it is agreed that Montagnier's group first isolated HIV, but Gallo's group is credited with demonstrating that the virus causes AIDS.[16] Since its discovery in 1981, AIDS has rivaled the worst epidemics in history. An estimated 25 million people have died, and 33 million more are living with HIV. Anti-retroviral drugs have been developed that can prolong the lives of patients.

THE EVOLVING IMAGE OF SCIENTISTS

Books that have popularized the scientific endeavor over the twentieth century to today trace the way idealism yields to self-interest and expediency, and collegiality and self-sacrifice are transformed into brutal competitiveness and arrogant ambition.

In Sinclair Lewis's 1925 *Arrowsmith,* an idealistic young doctor turned science researcher wrestles with temptations of success, money, and social status.[17] Ultimately, he finds a value system that satisfies his quest for intellectual and personal integrity. In a society of aggressive materialism, he resolves a conflict between science for commercial gain and science to benefit humanity. *Arrowsmith* was awarded the Pulitzer Prize.

Paul de Kruif's *The Microbe Hunters* followed a year later as a popular dramatization of the achievements of the pioneers in bacteriology.[18] Along with the Hollywood movies of the 1940s, such as *Dr. Ehrlich's Magic Bullet* and *Madame Curie*, these books were highly influential works depicting relentless dedication and personal sacrifice of scientists. César Milstein recalled that, as a schoolboy in Argentina, he was attracted to science after reading about early microbe hunters like Antoni van Leeuwenhoek and Louis Pasteur. Nobel laureate Rosalyn Yalow was inspired as a child by the scene in *Madame Curie* when Curie, after months of grueling toil, looks into her laboratory one night and finds her sample of radium glowing.

But with the 1968 publication of James Watson's autobiographical *The Double Helix,* the ego-driven striving for recognition and credit and raw competitiveness inherent in scientific research are ripped open and put on display.[19] While unraveling the structure of deoxyribonucleic acid (DNA) with Francis Crick at the Cavendish laboratory at Cambridge University in England is one of the most remarkable discoveries in history, Watson's telling of *how* it was discovered is just as astonishing. With the blunt portrayal of the admitted "shady" side of research, the reader experiences a loss of innocence. The benign portrayal of the selfless researcher is irretrievably altered. *Life* magazine's reviewer briefly noted that "the story should kill the myth" of scientific impersonality or the moral perfection of scientists: "These young scientists covet, lust, err, hunger, play and talk about it loud, well and long," and the *Chicago Sunday Sun-Times* noted that "what every scientist knows, but few will admit, is that the requirement for great success is great ambition. Moreover, the ambition is for personal triumph over other men, not merely over nature." But peering behind the curtain proved exciting. The book sold a million copies and was translated into almost twenty languages. Provocative and shocking are the tales of sneak peeks at other scientists' data, withheld information, and the no-holds-barred race against their chief rival, Linus Pauling.

They also record the joy of watching Pauling make a public blunder. Watson and Crick were indirectly provided a preliminary report that Linus Pauling planned to publish his proposal of a triple helical structure of DNA. They were delighted to find a major flaw in his concept, but, instead of warning him, they basked in his humiliation when the mistake was publicly discovered.

Rosalind Franklin, a fiercely independent expert in a technique using X-rays to picture the molecular structure of crystals such as DNA, at rival King's College in London, took numerous images of the DNA molecule. Observing the photos—without her permission or knowledge—provided Watson and Crick the triggering clue for the recognition of its double-helix structure. Of Franklin's meticulous series of X-ray pictures, one that would become widely known as Photograph 51 presented scrambled smudges to an uninitiated eye. But Watson, with his prepared mind, was struck by the symmetry of the DNA molecule: right-to-left and top-to-bottom, interpreted as a double helix, like a spiral staircase with two handrails.

Watson's account demonstrates the subtle undermining of Franklin's contribution to the discovery of the structure of DNA, as well as blatant discriminatory language characterizing her as a woman who does "not emphasize her feminine qualities…The thought could not be avoided that the best home for a feminist was in another person's lab." Franklin died of cancer in 1958 at the age of thirty-seven, four years before the Nobel Prize was awarded to Crick and Watson of Cambridge University and Maurice Wilkins of King's College.[20] The Nobel rules do not allow the award to be given posthumously.

Gary Taubes's aptly titled *Nobel Dreams: Power, Deceit and the Ultimate Experiment*, published in 1986, dispels any romantic visions of scientists as ivory tower innocents.[21] Carlo Rubbia, flamboyant and brilliant director of a 150-strong collaboration of physicists at the international accelerator laboratory at the European Center for Nuclear Research (CERN) near Geneva, won the 1984 Nobel Prize in physics. Rubbia and his team had discovered certain subatomic particles that are carriers of the so-called weak force that, together with electromagnetism, the strong nuclear force, and gravity, constitute the known fundamental interactions in nature.

The book suggests that Rubbia's overweening ambition, ferocious single-minded pursuit, and ruthless maneuvering make James Watson look like a choirboy. The book quotes a physicist who worked with Rubbia, saying that Rubbia obviously "read and assimilated Machiavelli." Rubbia drove himself and his co-workers mercilessly, and the book is an account of opportunism in big science, and of the frequently violent interactions between personalities and factions. The book suggests that Rubbia's unrestrained enthusiasm for a particular preconceived result even biased his analysis of the experiments. The scientists were all there because they were driven by the physics, but personal ambitions, power and glory, politics, and backbiting also came into play. The same compulsive desire that drives scientists to be the first to uncover nature's secrets can drive them to need maximum credit and public adulation.

Allegra Goodman's novel *Intuition* (2006) uncovers the sacrifice of principle for expediency in a high-profile medical research laboratory.[22] It dramatizes the tension between the pursuit of truth and the hunger for quick fame, the conflict between the painstaking pace of research and the pressure to produce results rapidly to obtain grant money. The lab's co-director—a brash, politically savvy oncologist—launches an accelerated PR campaign to promote a postdoctoral fellow's highly preliminary results of an anti-cancer experiment in animals. "We can wait," he persuades his cautious co-director, "until we've dotted every 'i' and crossed each 't' ... Or we can seize the moment now. We can announce results that are still preliminary. We can risk that [the results are] incorrect and stake our claim before someone else does." However, allegations of fraud by a whistle-blower lead to an investigation by the National Institutes of Health and hearings in Congress. Goodman's tale evokes the scandal Nobel laureate David Baltimore faced in the late 1980s, which I will describe more fully in the next chapter.

Thus, we can see the remarkable transformation in how the scientist has been depicted. Until fairly recently, he was apt to be viewed as a high-minded truth seeker. Watson's account first revealed the colorful personalities involved and the drama shaped by their relationships with one another. His opening sentence in *The Double Helix* is, "I have never seen Francis Crick in a modest mood." He further lets us know that he thought Rosalind Franklin was secretive and stubborn, that he himself—at the age of twenty-three—lusted to be famous. By the time of Taubes's and Goodman's books, the social behavior of scientists had become the center of attention. Human drama, underlying the intellectual pursuits, shows scientists have the characteristics shared by the rest of us—they are ambitious, competitive, and egotistical.

CHAPTER 4

THE DARK SIDE OF SCIENCE

The scientist who cheats on his research violates the basic precept of science, which is to find the truth and make it known.

—*Dr. Ned Feder, National Institutes of Health*

Starting in the 1970s, many in the United States have increasingly come to realize that major arenas of public life once thought sacrosanct can be seriously tainted by corruption and fraud. Centers of authority and social standing were confronted and their compromised integrity exposed. The government and military were openly challenged over the Vietnam War, the Watergate and Iran-Contra scandals, and the Iraq War. Corporate boardrooms sheltered conspiracies of massive fraud and cover up, from General Motors's denial of problems with the Corvair to tobacco companies' collusion to downplay the addictive properties of nicotine. The 2000s saw the bankruptcy of multiple corporations due to fraudulent accounting practices: Enron (the fifth largest corporation in the United States), Global Crossing, WorldCom, and Tyco International. Journalistic integrity was marred by several cases of blatant fabrication in the most prestigious media outlets. The Catholic Church has been plagued by sex abuse scandals it has denied and covered up for decades. Most recently, massive bank defaults from unethical transactions have led to the most serious recession since the Great Depression. Corruption appears pervasive among politicians and other public servants.

Accompanying and reinforcing the culture of fraud is the star-worshipping celebrity culture. Andy Warhol's famous observation that we can each attain fifteen minutes of fame was prescient. Those adept at the process of self-promotion may become famous for nothing more than being famous.

Fame, glory, and money are the goals of both the corporate and celebrity cultures.

Such instances prepare us for understanding how a foundation of trust can be easily subverted even in the world of science.

SCIENTIFIC FRAUD

As in any human enterprise, science as a discipline is not immune to fraud. How common is scientific misconduct? In a poll of two hundred British high-energy scientists, the journal *Science* noted that over a sixth believed that some of their ideas had been stolen by coworkers. Half of the American physicists who responded to a similar inquiry said that they would not feel comfortable discussing their ideas with all of their colleagues.[1]

In 2009, Daniele Fanelli analyzed survey data on how often scientists in the United States, the United Kingdom, and Australia engaged in misconduct in the past two decades.[2] On average, about 2 percent of scientists admitted to having fabricated, falsified, or altered results to improve the outcome—"a serious form of misconduct by any standard"—and up to one-third admitted a variety of other questionable research practices, including "dropping data points based on a gut feeling" and "changing the design, methodology or results of a study in response to pressures from a funding source." In surveys asking about the behavior of colleagues, similar misconduct had been observed, on average, by over 14 percent of respondents and other questionable practices by up to 72 percent. Indeed, since researchers may understandably be unwilling to reveal such actions despite all guarantees of anonymity and since the history of science shows that those responsible for misconduct have usually committed it more than once, the actual prevalence of biased or falsified data in the literature is likely underestimated.

More to the point, the Office of Research Integrity in the U.S. Public Health Service, Department of Health and Human Services, and the Office of Inspector General of the National Science Foundation investigate scores of misconduct allegations every year. Such investigations have led to the conviction of scientists, trainees, and technicians.

Scientific fraud is particularly dramatic because it is committed by privileged members of society whose basic tenet is truth. Scientists—venerated for breathtaking accomplishments like the Manhattan project, the polio vaccine, the discovery of antibiotics, cardiac surgery, and space travel—are human, too, and may yield to unethical actions. This is the dark side of science. Yet this truth may be particularly instructive for understanding science's varying methods, professional and social constructs, norms, and values. We can learn a lot about how science functions by illuminating its dysfunctional aspects. Indeed, examining the nature of fraud and other misconduct provides a way of reaching to the pulse of what scientists are and what scientists do.

This kind of misconduct can last and even prosper for a surprisingly long time, despite the oft-repeated claim that science monitors itself. The gatekeeping roles of a researcher's mentors, peer reviewers, journal editors, and sophisticated readership can simply be hoodwinked for a time, though invariably a misstep exposes the misconduct. Such a faux pas has been described as "the thirteenth chime which casts doubt on all that had preceded it."

While rare instances of alleged scientific misconduct date back several centuries, public disclosure of scandal in science and frequently frenetic news coverage first surfaced in the late 1970s and early 1980s. By 1980, advances were being made on the understanding and manipulation of genes, the nature of viral infection, and the mechanisms of cancer. Federal obligations for basic biomedical research had sharply risen to roughly $2 billion. In the early 1980s, for the first time, congressional hearings questioned the extent of fraud in biomedical research. In response, there arose definitions and rules about scientific misconduct, institutional policies for pursuing allegations, the teaching of research ethics and scientific integrity in graduate curricula, and policies of federal funding agencies, which have the force of law.

WHY DO SCIENTISTS CHEAT AND LIE?

The same impulses that lead someone to pursue scientific research and strive for recognition may, under certain circumstances, prove to be extremely dysfunctional and contrary to the integrity of the search for knowledge. The issue has been spotlighted over a span of years by science journalists, first by William Broad and Nicholas Wade[3] and more recently by Horace Freeland Judson,[4] authors who view such fraud through the lens of betrayal, a term embraced in their book titles. Nobel laureate Salvador Luria suggests that a peculiar pathology exists in the personality of one who would cheat in science.[5] Only a distorted sense of reality, he argues, could account for someone who would falsify or fabricate results. Thinking one could get away with such behavior in science, where external and internal control measures continually demand verification, would be a delusion. But this opinion does not take into full account the complexity and nuances of the dynamics involved. What motivates someone to commit scientific fraud? Many cases involve scientists at the start of their careers, like graduate students and postdoctoral fellows, or scientists in their initial career position, for example, assistant professors or low-rung corporate researchers.

Science is a highly demanding career-driven discipline. Researchers live in a world constantly short of grant money. With salaries dependent on grants and grants dependent on outcomes, the pressure to come up with winning experiments is intense. Since a good reputation relies largely on the publication of high-profile scientific papers, the imperative to "publish or perish" is a sword of Damocles. Many articles and presentations offer little substance to an original contribution, under the motto "publish a lot; polish a little." The number of scientific journals continues to increase each year, and the incessant grinding out of papers typically adds little of significance. The physicist Wolfgang Pauli, known for his acid tongue, once remarked to a colleague: "I don't mind your thinking slowly, but I do mind your publishing faster than you think."

Simply put, there is no journal of negative results—that is, research is generally publishable only if it reports positive outcomes. The more intensely the scientific enterprise values originality and recognition, the greater is the pressure for creative achievement, the distress over personal failure, and therefore the vulnerability to fabricate results. A researcher may be seduced by the siren call of fame, money, and envy to present research in its most promising light. He may believe in a hypothesis so strongly that data that do not seem to fit are edited out.

In Robert Millikan's famous "oil drop" experiments, which measured the charge on the electron—work central to final establishment of the atomic theory that led to his Nobel Prize in physics in 1923—he intuitively discounted data that did not conform to his expectations. Some have argued that he was simply exercising scientific judgment, but he is faulted for writing in his published work that he presented all of his available data on "all of the drops experimented on during 60 consecutive days." In his autobiography, he repeated this and added "no single [drop] being omitted." His notebooks, however, show that lots of drops were omitted. In this manner, to achieve scientific precision, Millikan reduced the statistical error of the charge on the electron from 2 percent to better than one half of 1 percent. This begs the question: Particularly if the results had not been validated, would Millikan have been charged with scientific misconduct?

Or the researcher may believe that he "knows" the answer and can take shortcuts to get there. With a deeply held personal conviction of being on the right path, despite glitches, scientists may cling to the thought that time will prove them right. In such cases, they may not view suppressing data as a guilty action. Further, the knowledge that some experiments yield data that are not precisely reproducible may be exploited. Is violation of a principle of scientific research excusable if it validates a truth?[6] The scientist who navigates these questions engages in a complex mix of self-sacrifice and guile.

Particularly in large-scale biomedical research groups, the team's junior scientists are typically distant in the hierarchy structure from the research czar (the principal investigator, PI) whose prestige has attracted grant funds. Although the PI's contribution to the actual work of his group may be nominal, he is credited on publications in the form of "honorary authorship." Because of his reputation and influence, the work carries a stamp of credibility and then becomes generally attributed to him. This has been designated by Robert Merton as the "Matthew effect." "For unto every one that hath shall be given, and he shall have abundance; but from him that hath not shall be taken away even that which he hath," says the text in St. Matthew's gospel.[7] This may create a resentful "hired-hand" mentality in the junior scientist, who may even be the first author on the paper. Reporting relationships may be unclear, accountability may be dispersed, and temptations to modify the results to fulfill the chief's expectations may be irresistible.[8] Out of pressure or resentment, or even as a subconscious challenge to his mentor to uncover the deception, the junior researcher may resort to misconduct.

IT'S IN THE AIR

A common apprehension among researchers is that there are other scientists who are close to success in the same experiment or theory development. Occasionally, fundamental concepts have ripened so that an astute researcher can readily harvest them. The idea is "in the air," social conditions and technology are both advanced enough to make a discovery absolutely inevitable. This puts extra pressure on researchers to be first. Throughout scientific history, as we've seen with Darwin and Wallace, similar discoveries have been made almost simultaneously by two or more people working independently. Especially in these circumstances, commentators resort to proclaiming the presence of a zeitgeist. Goethe first used the word in 1827 in referring to the unconscious, covert, and implicit effects of the climate of opinion. Zeitgeist, a German term with no equivalent in English, literally means time ghost or "the spirit of the times" and is used in this context to connote prevailing cultural and, especially, scientific opinions.

At the heart of such simultaneous discoveries is the apparent force of inevitability. Scientists not only know this but fear it, and this often activates a rush to ensure priority. "In contemporary science," states the main character in Carl Djerassi's *Cantor's Dilemma*, "the greatest occupational hazard is simultaneous discovery."[9] We see this in the rush to discover the structure of DNA. "Don't you see," Francis Crick told Anne Sayre, Rosalind Franklin's biographer, "if I hadn't [taken on the DNA problem], Pauling would have got it out first. I know Linus was wrong in his first guess, but Linus isn't stupid... he'd have done it." Crick seems equally certain that Rosalind Franklin would have solved the problem, "with Rosalind it was only a matter of time... perhaps three weeks, three months is likelier."[10]

Another striking example of simultaneous development is the telephone patent applications of Alexander Graham Bell and Elisha Gray, which arrived at the patent office within hours of each other on the same day in February 1876. Gray and Bell settled their differences, with Gray getting a $100,000 settlement and lucrative contract work. Bell's patent application was approved within weeks and became one of the most valuable in history.

In the same week in 1970, the journal *Nature* received separate papers from two of the world's most noted biologists: David Baltimore at MIT[11] and Howard Temin at the University of Wisconsin.[12] They had simultaneously and independently discovered a special enzyme, reverse transcriptase, that can transcribe genetic information carried in an RNA molecule back into DNA. Five years later, Baltimore and Temin shared the Nobel Prize. The discovery of reverse transcriptase made possible a key technology in genetic engineering.

SCIENTISTS BEHAVING BADLY

Scientists do make honest errors and have honest differences in interpretations of data, and these ought not to be confused with or judged as misconduct. Honest error from fanciful exuberance is illustrated by this example: At an 1858 meeting of the London Linnean Society, of which Charles Darwin was a prominent member, one member earnestly claimed that he had discovered fossil evidence of a flying dinosaur. Imprints of earth-bound creatures could be credibly accepted, but the society was intrigued by the proposition that there was evidence of a creature in flight. What was offered was the statement that the creature's wing tips, in ever-accelerating beats and a gradually uplifted motion, could be traced by ever more shallow imprints in the ground at the appropriate distance from each other. If one observes the common splashing at the tips of the beating wings of a water bird as it flies upward, the effect can be easily pictured. Nevertheless, the claim was promptly discredited as untrue.

However, defining and dealing with behavior that falls between honest error and fraud can be difficult. Three deviant activities constitute serious forms of scientific misconduct: fabrication (making up data or results), falsification (willfully distorting data or results), and plagiarism (copying ideas, data, or words without due acknowledgment).

Fabrication

William Summerlin, a dermatologist with a Ph.D. in immunopathology, had worked in the laboratory of Robert Good, a founder of modern immunology, for three years at the University of Minnesota. Summerlin's experiments with mice indicated a potential major breakthrough: When skin and other organs were kept in tissue culture for four to six weeks, they appeared to lose their normal ability to provoke an immune response. If this proved true, tissues could be grafted and transplanted between genetically mismatched individuals without rejection. A new era with vast promise dawned in the field of transplantation.

When Good moved to New York in January 1973 as president of the Sloan-Kettering Institute for Cancer Research, affiliated with Memorial Hospital, he recruited Summerlin, then thirty-five years old, as a full member of the Institute (the equivalent of being a full professor) and chief of transplantation immunology. Some critics felt that Summerlin lacked adequate training in research. Pressed initially with clinical duties, which included setting up Memorial Hospital's first dermatology unit, he resumed active laboratory work in the winter of 1974.

But Summerlin's career was about to collapse upon him. A research fellow working in his lab with the express mission of repeating the Minnesota mouse work simply could not confirm the findings, regardless of a variety of techniques. Further, Good heard that other investigators were holding off publishing negative reports out of respect for him.

Early one morning in March, Summerlin was summoned to Good's office to confer on the preparation of an article for the journal *Transplantation* on the negative results of the research fellow's efforts. Summerlin's name was to be included on that paper. Summerlin, who had idolized Good, was now resentful, feeling that his mentor was turning against him. Good was known for his aggressive style, so Summerlin took along eighteen white mice that had been grafted with patches of skin from black mice. In the privacy of the elevator as it ascended to Good's office on the thirteenth floor, Summerlin impulsively took his felt-tipped pen from the breast pocket of his lab coat and darkened the dull grayish graft areas on the skin of two of the mice. During the meeting, he showed them to Good as evidence of a successful graft. Good only glanced at them, focused instead on the consequences of the *Transplantation* paper.

Lab assistants, returning the animal cages to their places, noticed that patches had been drawn on the mice and could be wiped off with a ball of alcohol-soaked cotton. Summerlin then admitted to the paint job, and Good immediately suspended him. A committee within the Institute reviewed all of Summerlin's research, including his results in Minnesota, and found profound and serious "misrepresentation" (they never used the word "fraud") and a "lack of properly organized and analyzable data." Investigations revealed that earlier data concerning the success of human cornea transplants into rabbits was also highly suspect. Fabricated data had been cited in grant applications.

Summerlin later attributed his deceptive behavior to a combination of mental and physical exhaustion, a crushing workload, and pressure to publicize positive results and obtain grant funding. Lewis Thomas, a physician and administrator with wide experience (known to the general public as a graceful essayist), was the president of the entire Memorial Sloan-Kettering complex. He had been chairman of the department of pathology and subsequently dean at New York University Medical School and Yale Medical School. Thomas said that Summerlin was suffering from "a serious emotional disturbance." One skeptic thought that Good, who certainly bears some responsibility for promoting undue publicity surrounding Summerlin's claims without adequate authenticated data, "suffered from 'Nobelitis'." Two years earlier, Good wrote of Summerlin, "I am deeply impressed [by him]. I am certain he is honest to the core." Now he exclaimed, "I just don't know why it took me so long to disbelieve Summerlin."[13]

Musing on the episode, Summerlin speculated that perhaps his inking of the mice was a kind of challenge, a gesture to test Good's attention and acumen, as if to say, "If you're so smart, pick out the phony mice."[14] Foretelling even more complex scientific frauds yet to come, Jane Brody in the *New York Times* described the Sloan-Kettering incident as "a medical Watergate" that reflected "dangerous trends in current efforts to gain scientific acclaim and funds for research." She did not let Good off the hook, accusing him of "manipulating national attention and attracting an enormous amount of money for the Institute."[15]

That was the end of transplantation without immunosuppression. It sank without a trace in the literature. Summerlin is now reportedly a dermatologist in private practice. But the trouble doesn't end there. Investigation of alleged misconduct not only tarnishes the image of science and scientists but can also damage the reputation of the institution. The media reflexively referred to the Summerlin episode, for instance, as "the Sloan-Kettering affair."

Falsification

Falsification may convey not only conscious fraud but also nuances of bias, misguided interpretation, and simple carelessness, which can also cause truth to be misrepresented or misinterpreted, still with significant consequences.[16] Many researchers may unwittingly cross the fuzzy border between removing noise from results and biasing them toward a desired outcome.[17] But the intention to deceive—deliberate dishonesty—is a key element of all falsification. It is relatively easy for a researcher to cheat because the defined role of a reviewer is to judge whether the conclusions of a paper submitted to a journal are merited by the data, not to uncover concocted data.[18]

"Intellectually brilliant" was how those who came into contact or worked with John Roland Darsee characterized him. He was a respected research fellow at the Harvard Medical School in the laboratory of the cardiologist Eugene Braunwald, holder of one of the most distinguished chairs of medicine in the nation and a member of the National Academy of Sciences. Braunwald's two-volume, 2000-page *Heart Disease: A Textbook of Cardiovascular Medicine,* published in 1980, became an instant classic. Darsee was co-author of a chapter. Aged fifty-two, Braunwald's name was on over 600 papers, and the research empire he oversaw was vast. This involved two laboratories, one at Brigham and Women's Hospital, an affiliate of the Harvard Medical School, where he was physician in chief, the other at Beth Israel Hospital, a couple of blocks away. Almost twenty research scientists worked on projects funded with $3.3 million in NIH grants. It was apparent to some that Braunwald was grooming Darsee to head the lab at Beth Israel Hospital.

By the age of thirty-three, Darsee had published more than one hundred papers and abstracts over a two-year period, many of them co-authored with his mentor. Darsee was a major participant in a National Institutes of Health (NIH)-funded study designed to assess whether certain drugs limit the damage done by heart attacks.

To some, Darsee's prodigious publications and rapid ascent in the biomedical field were suspect. It was seriously questioned whether an abstract he was preparing for publication was based on any actual research. Challenged, Darsee could not provide the experimental raw data. Rather, in May 1981, as observed by several colleagues in the lab, Darsee simply started to collect and label randomly acquired "evidence." The act was so brazen that Darsee had no choice but to admit to the falsification, rationalizing that he had discarded the real raw data. He denied any other improprieties. An audit by Harvard revealed that the original data for many of Darsee's experiments were unaccountably missing.

Braunwald urged against measures that would tarnish Darsee's bright future, and for six months Harvard authorities took no public action. In fact, Darsee was allowed to continue working in the laboratory on further experiments, including a $724,154 multi-institutional study sponsored by the NIH. Braunwald's wish to "manage" the situation suggested to some a cover-up. The resultant delay stands in contrast to Robert Good, who had suspended Summerlin the day he confessed. No notice was given to Darsee's many collaborators that they were at risk, even though several had contributed little, serving mainly as honorary co-authors. Yet, it became evident that the period in which the volume of raw data started to fall off was after Darsee had been at the lab for a little over a year—that is, the point at which supervision relaxed. Braunwald pointed out that after some time, "it is not the custom to hold onto a fellow's hand at every turn." Nevertheless, one faculty member of the Harvard biochemistry department expressed shock at Braunwald's lack of responsibility: "The idea of [the NIH] coming down to see the original data, implying that papers had been sent off based on work the senior researcher had never looked at, is quite amazing to me."[19]

Harvard finally acted. It withdrew Darsee's results from the study, released a brief press statement about the admitted falsification, and appointed a committee of five professors to investigate and make recommendations. They concluded there was no evidence of fraud. However, a committee set up by the NIH reported a radically different conclusion in early 1983: Almost every paper produced by Darsee was fabricated. Its report further criticized Harvard's "management" of the investigation as well as the lax supervision and poor record keeping in the lab, which was under Braunwald's direction. Later, duplication of data in separate research publications and publication of the same abstract in multiple journals were uncovered. Many of Darsee's more than one hundred studies proved to be fraudulent and had to be retracted.

Darsee lost his Harvard position and was banned from NIH research grants for ten years. Harvard repaid the grant money.[20]

Plagiarism

The term *plagiarism* derives from *plagiarius*, "to kidnap," and it signifies breaking a connection between an author's name and their work. In 1980, this misconduct led to one of the most serious shake-ups in the history of biomedical research. It involved two premier medical centers and the NIH, congressional hearings, public dismay over excesses of scientific fraud, the discharge of an eminent researcher and administrator, and a reformulation of the need for investigation of alleged scientific misconduct and for definitions of the responsibility of authorship.[21]

In this famous case, a whistle-blower's charge of plagiarism was the opening salvo that led to a confession of data fabrication by a thirty-seven-year-old assistant professor at Yale Medical School, Vijay Soman. His published work, which dealt with aspects of insulin in patients with anorexia nervosa, had been performed under the supposed supervision and co-authorship of Philip Felig, a forty-three-year-old distinguished professor with more than 200 publications and vice-chairman of the department of medicine at Yale. The case took on Shakespearean tones as it was revealed that both Felig and Soman negatively reviewed a rival's manuscript that had been submitted to the *New England Journal of Medicine* rather than recusing themselves on the basis that Soman was privately working on an identical study. Soman, using information from the rejected manuscript, shortly mailed off a paper on the subject to the *American Journal of Medicine*, where Felig was an associate editor. By a remarkable twist of fate, the paper landed for review on the desk of Helena Wachslicht-Rodbard at the NIH, the author of the rejected manuscript. An outraged whistle-blower was born. She demanded an investigation by Yale, not only for plagiarism, including of a formula she had devised, but also for apparent manipulation of data. In explaining the situation to a reporter, Yale officials at times referred to Rodbard as a "hysterical female."

Fate had yet another hand to play. In January 1980, Felig accepted an offer from Columbia University's prestigious College of Physicians and Surgeons to become the Samuel Bard professor and chairman of the department of medicine. He introduced Soman to officials at Columbia and recommended that he be appointed assistant professor. Felig planned to start in June 1980.

Yale delayed an investigation for nearly one year, and then in the course of a scientific audit in February 1980 that exposed inadequate and fabricated data, Soman confessed to modeling his paper on Rodbard's and to falsifying work. "[My] actions," he told an auditor, "were done in the midst of significant pressure to publish these data as fast as possible so as to obtain priority." He agreed to resign and to leave Yale; he returned to his native India. Further investigation determined that many other published articles evidenced missing data or outright fraud. Ten of these papers were co-authored by Felig. Eventually a dozen papers were retracted.

A little more than two months after assuming his post at Columbia, a six-member committee at the medical school forced Felig to resign, in part because he had "failed to communicate" the allegations of plagiarism and because of "his failure to grasp the significance of incriminating facts." Other charges included "poor judgment," "character assassination," "ethical insensitivity," and "unacceptable standards."[22]

In a fall from grace, Philip Felig, after having his case reviewed by the Yale Medical School for more than three months, was rehired, but he lost his endowed chair and was discontinued from serving as associate editor of the *American Journal of Medicine*. He exposed a flaw in his perception of responsibility regarding Soman's research publications. "In essence," he told the congressional subcommittee on investigation and oversight, "I was providing Soman a cloak of scientific credibility." Yale subsequently established administrative mechanisms to investigate charges of misconduct and spelled out the responsibilities involved in team research: "Each collaborator must accept the discredit, as well as the credit, for the collaborative effort."

And what of the whistleblower, Helena Wachslicht-Rodbard? In April 1979, her paper was published in the *New England Journal of Medicine* in revised form. But, feeling that research held no attraction for her, she left the NIH and entered private practice.

Probably the most common type of scientific misconduct is citation plagiarism, willful or negligent failure to appropriately credit other or prior discoverers in order to give an improper impression of priority. It may be difficult to judge in any instance whether this is intentional or simply based upon ignorance of the prior work. Cryptomnesia, or unconscious plagiarism, may be offered as a defense based on an author's claim of having absorbed some fact or idea and honestly believing it was his invention.[23] Memory is mistaken for imagination, because there is no subjective experience of it being a memory. The German language has a term for this that translates to, "writing with the left hand."

Quite another perspective was taken by T. S. Eliot. In the humanities, ethical constraints and the criteria for originality may be less strict. "The Waste Land," a haunting statement of his disillusionment with the post-war era, is still celebrated as one of the world's greatest works of poetry. Much of the poem, however, has been shown to be cobbled together out of quotes from other writers. Eliot's explanation: "Immature poets imitate; mature poets steal." At first blush, this may seem an arrogant assertion of the artist's right to be "influenced" or "inspired" by a previous work, but is it so different from the fundamentals of an act of creation as described by Arthur Koestler? "The creative act is not an act of creation in the sense of the Old Testament. It does not create something out of nothing; it uncovers, selects, re-shuffles, combines, synthesizes already existing facts, ideas, faculties, skills. The more familiar the parts, the more striking the new whole."[24]

Synthesis, more than analysis, is the key cognitive skill. Creative thinkers tend to bring synthesis to higher levels. They have a gift for seeing *similar differences* and *different similarities*—phrases coined by the British theoretical physicist David Bohm. True creation, Bohm argues, relies upon perceiving a new fundamental set of similar differences that constitutes a genuinely new order.[25]

The great conductor Arturo Toscanini once encountered a form of reverse plagiarism—credit falsely advertised. One evening as he left La Scala in Milan, where he had conducted *The Magic Flute*, he came upon an organ-grinder in the piazza. Toscanini, upset by the monotonous sound from his barrel-organ, berated him for not varying the speed of the handle and the tone of his instrument. The next evening, when Toscanini left the theater, the man was there again and the same monotonous sound filled the piazza. There was, however, one difference. A large sign behind the man now proclaimed: "Pupil of the great Toscanini."[26]

THE POLITICS OF SCIENTIFIC FRAUD

The most notorious case of alleged misconduct in basic medical research has become known as the "Baltimore Affair," even though the research contribution of David Baltimore, the report's most notable co-author and the fiasco's namesake, was itself never questioned.[27] Baltimore, who had been awarded the Nobel Prize in 1975 at the age of thirty-seven for his discovery of an enzyme that plays a key role in propagating some cancer viruses, built the highly respected Whitehead Institute for biomedical research at the Massachusetts Institute of Technology (MIT).

A dispute began in 1986 at MIT when Margot O'Toole, a thirty-six-year-old research assistant in her sixth year as a postdoctoral fellow, accused a senior scientist, Thereza Imanishi-Kari, in a detailed statement, of misreporting data in a paper published in the journal *Cell* in April that year, dealing with how immune responses are regulated. O'Toole was a competent researcher. Unable to repeat several of the basic findings in the paper, she concluded that the original laboratory data did not support the authors' published contention and argued for a retraction of the paper. She was careful to neither allege fraud nor use the word "misrepresentation." The work was done in separate labs in collaboration with David Baltimore, and he was listed as one of six co-authors. The research was supported by significant federal funding.

Baltimore went to surprising lengths to vigorously defend Imanishi-Kari and to shut out scrutiny as the incident escalated into a highly celebrated case of alleged scientific fraud. His obstinate handling of the case with characteristic self-confidence, which some of his colleagues called arrogance, drew intense criticism. In recognizing the ambiguity of scientific research, he staunchly underscored the scientist's need to make judgments about data. "A paper does not claim to be an absolute assurance of truth," he explained, "only a moment's best guess by one group of investigators."[28]

The matter drew two university reviews—from MIT and Tufts University Medical School, which subsequently hired Imanishi-Kari—both drawing favorable conclusions; a prolonged congressional inquiry by John D. Dingell, chairman of the House Subcommittee on Oversight and Investigations; and two formal investigations by the NIH, which is charged with investigating allegations of misconduct in federally funded research at universities. Dingell and his fellow reformers sought to establish accountability, whereas Baltimore considered government interference as poisoning the self-correcting nature of science. One side saw a national crisis of trust in publicly funded research, and the other side mobilized to keep the government out of the laboratories and leave science to the scientists. An NIH panel, while validating the science of the papers and clearing the authors of misconduct, nevertheless uncovered a number of "significant errors of misstatement and omissions." By her own admission, Imanishi-Kari was haphazard with record keeping.

In mid-1988, Dingell held a Kafkaesque hearing over the possibility of "fraud" without ever calling Imanishi-Kari or Baltimore. Baltimore then sent out a nine-page "Dear Colleague" letter to 400 scientists, warning of the threat of "the introduction of new laws and regulations that I believe could cripple American science." A massive response depicted Dingell's inquiry as a political assault on the foundations of science. *The Wall Street Journal* referred to the threat of a "science police." Even Dingell's hometown newspaper, the *Detroit News*, titled a searing editorial "Dingell's New Galileo Trial."

A reconstituted NIH panel concluded in February 1989 that there was "no evidence of fraud, manipulation or misrepresentation of data." Beyond this statement, however, James B. Wyngaarden, director of the NIH, chastised the article's authors for never meeting "to consider seriously the allegations or to re-examine the data to determine whether there might be some basis for the allegation."

In the fall of 1989, Baltimore was appointed as president of Rockefeller University in New York City. He immediately faced a press announcement against the appointment signed by more than a third of Rockefeller's full professors. In a fall from grace, Baltimore learned the virtues of a more conciliatory approach.

At Dingell's request, the Secret Service even employed forensic techniques for analysis of evidence. This sent shudders through the scientific community. An NIH investigation reported in May 1991 that some of Imanishi-Kari's data was falsified and/or fabricated. Further, it stated that Baltimore's prolonged and vocal insistence on dismissing Margot O'Toole's charges had delayed getting at the truth. A combination of fear and respect for the academic hierarchy and the people who controlled the funding for the labs made it too risky for other scientists to come forward and support O'Toole. Baltimore issued a mea culpa in a letter to the NIH[29] and a statement commending O'Toole for her "courage and determination" and expressing regret for "my failure to act vigorously enough in my investigation of her doubts." O'Toole felt she was blackballed from working in a science lab for several years but then found a research position in a biotechnology company.

Many likened the case to a Greek tragedy: Baltimore's character flaw led him to commit a sequence of unnecessarily aggressive acts defending what simply were probably errors in publication and thereby did damage to his own personal reputation. The entire three-year affair could have been nipped in the bud if he hadn't belittled O'Toole's reservations, attacked her as a "discontented post-doc," and refused to publish a mere correction letter. Dingell would likely not have held hearings on the case, and the damage to Baltimore's reputation would have been nil.

Baltimore had been viewed before these events as a role model for thousands of scientists throughout the world. Afterward, his public image suffered considerably. He retracted the *Cell* paper, with Imanishi-Kari withholding her approval, and after eighteen months at Rockefeller University, resigned from his position to return to MIT.

In June 1996, a painful period of ten years after the initial charges, Imanishi-Kari was officially exonerated by the U.S. Department of Health and Human Services.[30] This act, in the minds of many scientists, also vindicated Baltimore's adamant behavior. The case now represents for many the worst aspects of government handling of scientific misconduct allegations. The following year, David Baltimore was appointed president of the California Institute of Technology in Pasadena, and in 2006 he was elected president of the AAAS.

HOAX!

Instances of blatant deception are often referred to as hoaxes, thus suggesting that the intention behind the offense was simply to pull an innocent prank. But can the swindled victims of Bernard Madoff's Ponzi scheme be persuaded that it was "only" a hoax? Neither can a hoax in science be treated lightly; it may bear dire consequences.

A famous paleontological hoax is the "Piltdown Man." In 1912, Charles Dawson, a member of the Geological Society of London, collected fragments of a human-like cranium and ape-like jaw from a gravel pit at Piltdown, a village in East Sussex, England. The discovery was hailed as the evolutionary "missing link" that supported the notion then prevailing that brain size increased first in the evolution from ape to modern man before the jaw adapted to new forms of food. Additionally, the discovery was enthusiastically welcomed as a British "dawn man" and embraced with nationalism and cultural prejudice. However, in 1953, it was exposed as a forgery—the bones consisted of a human skull of medieval age, the 500-year-old lower jaw of an orangutan, and chimpanzee fossil teeth. The appearance of age had been created by staining the bones with chemicals, and the teeth had been filed to give them a shape more suited to a human diet.

The forger clearly appears to be Charles Dawson, whose most cherished goal—that of a fellowship in the prestigious Royal Society—ultimately went unfulfilled. Dawson died unexpectedly in 1916. The truth took forty years to emerge.[31] The fraud had a significant impact on early research on human evolution, which was thrown off track for decades. The prominent paleontologist Arthur Smith Woodward spent time at Piltdown each year trying to find more Piltdown Man remains until he died. In time, with the discovery of humanlike fossils in Africa, a very different pattern of evolution emerged. The African fossils revealed that our predecessors actually had human-like jaws with ape-like skulls. Science may at times allow itself to be easily misled if findings are molded to confirm a bias.

PART II

CHAPTER 5

"DROP EVERYTHING!"

What has been achieved is but the first step; we still stand in the presence of riddles, but not without hope of solving them. And riddles with the hope of solution—What more can a man of science desire?
—Hans Spemann, Croonian Lecture, 1927

"Drop everything!" Selman Waksman excitedly rushed to his laboratory and with these words to his assistant, Boyd Woodruff, announced that he was altering the course of his research. Waksman was the distinguished soil microbiologist at the New Jersey Agricultural Experiment Station of Rutgers University in the city of New Brunswick. "See what those Englishmen have discovered a mold can do. I know the actinomycetes will do better!"[1] It was September 1939, and he had just returned from the Third International Congress for Microbiology in New York City, where he had met Alexander Fleming and had learned of Howard Florey's and Ernst Chain's work at Oxford University proving penicillin's therapeutic potential.

Fleming's accidental discovery of penicillin made real Louis Pasteur's idea that there are naturally occurring antibiotics. In 1928 at St. Mary's Hospital in London, Fleming noticed that a petri dish on which he was growing dense colonies of bacteria was contaminated by an airborne mold. The airborne mold astonishingly dissolved the bacteria. The finding languished for eleven years until the "mold juice," labeled penicillin, was isolated, purified, and clinically tested by the Oxford team.[2]

Ever since Pasteur discovered that many of man's most dreaded diseases are caused by microorganisms, scientists had searched for a drug that would kill them without damaging human tissues. A few chemical drugs were synthesized. Salvarsan—an arsenic compound—also casually known as "606," developed by Paul Ehrlich in 1909, proved to be effective against syphilis. Much later, in 1935, came the sulfa drugs, the medical wonders of their day, derived from industrial dyes. These shattered a myth of long standing: the invulnerability of bacteria. But none of the chemical "magic bullets" was effective against more than a few disease organisms, and all of them were apt to have toxic effects on human tissues.

Penicillin, a product of nature, would prove to be a revolutionary breakthrough in the treatment of infections. Although it was useless against tuberculosis, it worked wonders against many different types of bacteria. A scratch that could turn into a life-threatening streptococcal or staphylococcal infection was now treatable with penicillin, along with conditions such as puerperal sepsis ("childbirth fever") or pneumonia.

At fifty-one years of age, Waksman's abrupt reversal was not a mid-life crisis, but rather the realization that he was uniquely suited to further the breakthrough against the scourge of infectious diseases. Waksman (rhymes with "boxman") had attained recognition as an international authority in what he established as a scientific discipline, soil microbiology, and realized that the knowledge and, above all, the techniques of studying the soil's microscopic inhabitants could now be exploited for uncovering antibiotics for human use. This was, he knew, his moment.

HUMBLE BEGINNINGS

As a Jew fleeing Tsarist Russia in 1910—one of 187,000 Russian immigrants to arrive in the United States that year[3]—the twenty-two-year-old Waksman barely spoke English. Despite this, he promptly set about pursuing an education. There were obstacles: Not only did private colleges cost too much, but admission decisions were constrained by preference and prejudice. In certain scientific fields, furthermore, as delicately put by a historian, "attitudes unfavorable to people of Waksman's background prevailed."[4] But an opportunity for public higher education came about through the Hatch Act of 1887, a systematic program to establish agricultural experimental stations. The legislation specified that every state should have at least one such station. As the United States continuously expanded its frontier, agriculture was the one area of scientific research for which Congress generously appropriated money. From the 1890s to the 1930s, the Department of Agriculture

Figure 5.1   Selman Waksman's immigration photo. [WA]

was the leading agency of the federal government with scientific interests and health-related scientific work.

Family members directed Waksman to the Rutgers Agricultural College in New Brunswick, New Jersey. Dr. Jacob Lipman, a bacteriologist and a fellow Russian immigrant, headed the Agricultural Experimental Station and would soon be named dean of the College of Agriculture. Lipman, who would become a mentor, convinced Waksman that he should study agriculture at Rutgers and offered a scholarship if he could pass the exam in English—which he did. New Jersey's many truck farms, orchards, and flower gardens supplying Eastern cities give it the nickname the Garden State. As a student, Waksman flourished in his laboratory courses and became fascinated by the life teeming deep under the ground's surface. Every pinch of dirt contains millions and millions of microbes.

He dug a series of deep trenches at different parts of the College Farm, and, like a geologist who can unravel the mysteries of layers of different rock, Waksman studied the varying microbial inhabitants of soil depths. He counted the colonies, or groups, of bacteria and fungi and, to his surprise, came upon a third group of microorganisms. This was his first encounter with what would become his life's passion: a group of tiny microbes called actinomycetes, occupying a twilight evolutionary zone between molds and bacteria. Although these microbes are true bacteria, they closely resemble fungi in that they are filamentous and branching and reproduce by spores rather than by simple fission. Discovered in 1875, the actinomycetes were recognized as a group in 1890 but had been almost entirely neglected in previous research.

The young student observed an interesting finding. When he tabulated his year-end results, he discovered that the proportion of actinomycetes to bacteria gradually increased as he moved downward through layers of soil. What, he wondered, was the relationship between these microbes? How did the actinomycetes interfere with the population of bacteria?

Little did he suspect that actinomycetes would prove to be the main source of most of the clinically useful antibiotics. Twenty-five years would pass before his investigations would intersect with medical research and unlock the mystery behind those questions. His quest would bring him the world's honors but also the blackest day of his life.

He would later recall the "earthy" fragrance of the rich black dirt in his native Ukraine and how he came to understand that the smell was not of the earth itself but rather of its actinomycetes. The land was extremely fertile and, as he would later write in his autobiography, was "flat [with] wide, forestless acres. In summer, the fields of wheat, rye, barley, and oats formed an endless sea. In winter, snow covered the ground, and the frosted rivers carried the eye to the boundless horizon, where the skies met the earth somewhere far away."[5]

By the time he graduated in 1915 with a B.S. in agriculture, Waksman was sure that his strongest interest was in soil microbiology. He decided he needed to round out specific knowledge and spent a year as a research assistant in bacteriology in Dr. Lipman's department, earning a master of science degree. The year culminated in a further point of pride: American citizenship. Named originally after Solomon, he now transformed his first name to Selman. At this point, Waksman made another clear-headed decision: with an increasing awareness that even the smallest living cells have their own complexities, he needed to understand their underlying chemical interactions. With the support of a fellowship, he earned a Ph.D. in biochemistry at the University of California, Berkeley, in 1918.

By the age of thirty, Waksman had embraced soil biology, microbiology, bacteriology, and biochemistry, including enzyme chemistry. Experience in industry, at the Cutter Biological Laboratories in California at the time of his Ph.D. and subsequently at the Takamine Laboratory in nearby Clifton, New Jersey, testing the effects of drugs on the animal body, rounded out the picture. "I had sent my roots into the soil," he remembered in his autobiography. "I was on my way. I knew now exactly what I wanted and how to get it."

Waksman returned to work as a soil microbiologist at the New Jersey Agricultural Experiment Station, where he would spend his entire scientific life. He was a man with a mission. He now had a solid grasp of microorganisms and their life processes. His aim was to find practical applications of his science, particularly in improving soil fertility and crop production, and he knew that the organic chemical content of soil was a crucial factor. This appeared to be dependent on the proportionate populations of various protozoa, bacteria, fungi, and Waksman's favorite, actinomycetes. He analyzed soil, humus, peat, clay, loam, and sand for their microbes to see how they grew, how they multiplied, what nutrients they took in, and what waste products they exuded.

Such microbes exist through a delicate ecological competition, some producing chemicals to kill others. But still unknown was their potential value to modern medicine.

Over the years he traveled widely and carried out studies of peat bogs and composts throughout the United States, Europe, and the Middle East. He served as a consultant to commercial agricultural companies and many industrial concerns that produced enzymes, vitamins, and other products from fungal and bacterial sources. He was thus a forerunner of the entrepreneurs in today's highly developed biotechnology industry.

THE FERMENT OF DISCOVERY

The field of bacteriology at the time of World War I was divided generally into three areas: (1) the study of microorganisms important in food processing (the making of bread, cheese, wine, beer, and vinegar) and food spoilage. Some of the more than three hundred strains of the mold penicillium had been used by the French cheese industry for centuries. The blue veins of Roquefort cheese are due to the mold *Penicillium roqueforti*, and the distinctive taste of Camembert is imparted by *Penicillium camemberti*; (2) the study of pathogens by medical bacteriologists, which included sanitation problems such as water purification and investigation of sewage disposal; and (3) the study of microorganisms of concern to agriculture, including plant pathogens and those soil-enriching organisms typified by the nitrogen-fixing microbes.

Tantalizing observations on the phenomenon of microbial antagonism—substances produced by some microorganisms that are antagonistic to the growth of others—had been made for some time, but its therapeutic importance was simply not fully grasped. Pasteur and Joubert first noted it in 1877, observing in a famous phrase that "life hinders life."[6] They speculated on the clinical potential of microbial products as therapeutic agents, and ten years later another scientist hopefully commented, "Bacteriotherapy...now no longer appears to be in the realm of dreams as a means of fighting already developed disease."[7]

Yet for decades it was immunotherapy—vaccines and antitoxins—in Europe and in England—not direct bacterial antagonism—that dominated the therapeutic approach to infectious diseases. This was due not only to the influence of Pasteur's success with vaccines against cholera and rabies but in large part to the fundamental fact that all biological processes are chemically based and mediated, and thus progress in medicine often awaits progress in chemistry.

Just a year before Fleming's publication in 1928, a book published in France dealt extensively with bacterial inhibition by molds and by other bacteria, citing in a sixty-page chapter the concept of "antibiosis."[8] The word had been coined in 1889 as a counterpoint to symbiosis to emphasize the Darwinian struggle for existence.[9]

In 1924, Waksman traveled to Europe on a self-described "grand scientific tour," the first of many trips abroad, introducing himself to scientists at centers of soil microbiology to trade knowledge. He was persistent.

> I tried to present my interests everywhere I could. Often I was rebuffed, but more often I was received graciously and listened to attentively. Usually without proper letters of introduction, I would knock at the door of a famous scientist. Once I would gain admission, the interview would last much longer than one would anticipate. Only twice was admission refused me on the basis of improper introduction. And even in those cases, I was able to impress the person involved once I met him for a minute or two that my problem was important and that a more prolonged interview was in order.[10]

He was struck by the fact that they were all very familiar with bacteria and fungi, but no one knew about actinomycetes.

Within three years, the thirty-nine-year-old Waksman established himself as the world authority in his science with an 894-page textbook, *Principles of Soil Microbiology*, which gave the field the stature of a scientific discipline.[11] He attracted dozens of graduate students—not only from the United States but also from Canada, China, India, South America, and Europe—to work in his laboratory and study under him, helping to transform Rutgers from a

small agricultural college into a world-class institution. The scores of publications that issued from the Rutgers lab were possible due only to Waksman's
strong organizational abilities and tight control over the research products of
his students and associates. He was appointed to full professorship at Rutgers
and became a scientist of international distinction, a member of the French
Academy of Sciences, and, shortly, a member of the U.S. National Academy
of Sciences.

In physical appearance, Waksman was not an imposing figure. A short
stocky man with wire-rimmed spectacles and a thick moustache, he was indifferent to personal grooming. His white laboratory coat was often torn at the
elbow (ever practical, his attitude was simply "it serves the purpose"), and
otherwise he appeared in shapeless suits, often with food stains on his vest.
High-button shoes did not add to his sartorial splendor. But immediately
in conversation it would become evident that he could not be cast in the
role of absent-minded professor. He spoke with a trace of a Russian accent
but was otherwise precise in speech, having mastered the English language.
In fact, the introductions to his scientific books are notable for their lyrical
quality. In person, he was unpretentious and had an old-world courtesy of
manner. Recounting his efforts and accomplishments, he referred to "we" or
"my assistants and I" and never "I." He was warmly admired by his students
and assistants for his inspiring teaching and mentorship. It says much for
Waksman's character that though he himself was highly self-disciplined and
productive, his relations with his associates and in particular with his graduate
students and postdoctoral fellows were marked by patience and generosity.
When it came time for a graduate student or postdoctoral fellow to choose a
research project, Waksman would skillfully lead the discussion toward one in
keeping with the laboratory's interest. Work in his laboratory was done in a
spirit of enthusiasm and dedication. Many of his students went on to responsible careers in industry and academia.

OPPORTUNITIES MISSED AND SEIZED

Waksman missed several opportunities to make the great discovery earlier in his career, but his single-mindedness did not allow for, in Salvador
Luria's phrase, "the chance observation falling on the receptive eye." *I have*

the answer. What is the question? Turning an observation inside out, seeking the problem that fits the answer, is the essence of creative discovery. In 1975, Waksman recalled that he first brushed past an antibiotic as early as 1923 when he observed that "certain actinomycetes produce substances toxic to bacteria," since "around an actinomycetes colony upon a plate a zone is formed free from fungous and bacterial growth." Such a clear zone, a zone of inhibition on a Petri dish, is the hallmark of an antibiotic being secreted. Yet, Waksman acknowledged, "I paid only little attention to the significance of this phenomenon."[12]

Like Fleming with penicillin in 1928, Waksman was not prepared to pursue this inadvertent finding. From the mid-1920s through the 1930s, he brushed past one opportunity after another. In his 1926 book *Enzymes*, he briefly noted that "some bacteria produce enzymes which possess bacteriolytic properties,"[13] and in the 1932 second edition of his monumental *Soil Microbiology*, he observed that actinomycetes and fungi form "toxic substances which are injurious" to other microorganisms.

With opportunities to work on the chemotherapy of tuberculosis, he side-stepped each one of these opportunities.

In 1935, Chester Rhines, a graduate student of Waksman's, found that tubercle bacilli not only persist but grow abundantly in soil among a multitude of microbes. An incidental finding was that they would not grow in the presence of an unidentified soil fungus,[14] but Waksman did not think that this lead was worth pursuing: "My general impression at that time was that these results seemed to lead nowhere…In the scientific climate of the time, the result did not suggest any practical application for treatment of tuberculosis."[15] The same year, Waksman's friend Fred Beaudette, the poultry pathologist at Rutgers, brought him an agar tube with a culture of avian tubercle bacilli killed by a contaminant fungus growing on top of them. Again, Waksman was not interested: "I was not moved to jump to the logical conclusion and direct my efforts accordingly…I did not take advantage of the observation. My major interest at that time was the subject of organic matter decomposition and the interrelationships among soil microorganisms responsible for this process."[16] And yet, mindful of Pasteur's famous maxim that "chance favors only the prepared mind," he frankly acknowledged years later, "There also was undoubtedly some mental unpreparedness on my part."[17]

In 1942, his son, then a medical student, urged him to isolate strains of actinomycetes active against human tubercle bacilli, but he replied, "The time has not come yet. We are not quite prepared to undertake this problem. But we are rapidly approaching it."[18]

Today, we recognize that the ideal scientific mind comfortably incorporates unanticipated factors into an established body of work or, more likely, follows them in completely new directions.[19] In other words, the open mind embraces unexpected findings and converts a stumbling block into a stepping-stone. As Winston Churchill whimsically observed, "Men occasionally stumble across the truth, but most of them pick themselves up and hurry off as if nothing happened." It is facile to decry Waksman's apparent inertia, but it must be remembered that he was a renowned soil microbiologist, not a physician or a *medical* bacteriologist like Fleming, and his career was based on the quest for improving soil fertility. His fixed paradigm of thinking simply kept him from carrying the pattern of observations to a practical conclusion.

As inspiring as the International Congress for Microbiology in New York was for Waksman, he had already been primed to the potential of microbial antibiotics to act against human pathogens by the recent work of René Dubos, a former student in his laboratory. A tall Frenchman with thick glasses and a robust personality, Dubos had a piercing intelligence. Working side by side with medical researchers at The Rockefeller Institute for Medical Research in New York City (renamed The Rockefeller University in 1965), Dubos reasoned that as all organic matter eventually breaks down in soil, soil microbes must be able to dismantle or destroy cells of all kinds—including those of dangerous bacteria.

But Dubos's approach was highly original. Rather than using enriched culture media, as other scientific investigators did to favor the growth of microorganisms, he restricted the nutrients for growth, encouraging the potential of a bacterial cell to produce an enzyme to carry out the decomposition of the substance it was attacking. It was a fantastic idea: to feed disease germs to soil microorganisms and see which species thrived on the diet.

The standard method of differential staining of bacteria was discovered decades earlier by a Danish physician, Hans Christian Gram. He discovered that in a small sample of body fluids—for example, saliva—stained with dyes, microscopic examination could distinguish different forms of bacteria.[20]

There are three shapes of bacteria, serpentine, round, and rod, and they differ in the kinds of illnesses they produce. The world of bacteriology has since then been divided into gram-positive and gram-negative organisms. Some bacteria stain blue; they are called gram-positive. Others stain pink; these are gram-negative. Gram-positive and gram-negative differ in the kind of infections they cause. The two types, as would be shown, react differently to antibiotics. The sulfa compounds, for example, were useful chiefly against the gram-positives.

In general, gram-negative bacteria are more resistant to antibiotics because their outer membrane prevents drug penetration. After two years of work, with the help of Rollin Hotchkiss, an organic chemist, Dubos isolated a substance he dubbed gramicidin, lethal to streptococci, staphylococci, and pneumococci.[21] He derived it, using his own novel method, from a common soil bacterium, *Bacilli brevis*. It was, however, too toxic for human use except as applied externally in the form of commercial ointments or salves for infected ulcers and wounds. It achieved some popularity when successfully used for an udder infection, in Elsie, the Borden cow at the 1939 World's Fair. Nevertheless, the threshold to a new era had been crossed.

Dubos alerted scientists that it was possible to find soil-based bacteria that inhibited the growth of disease-causing infectious bacteria. This was the first antibacterial agent to be obtained from natural sources, through rational pursuit. In this way, Dubos touched off the antibiotic revolution. He reported on gramicidin with a showman's brio at the Third International Congress of Microbiology, which opened at the Waldorf-Astoria in New York on September 2, 1939, the day after Hitler invaded Poland and the day before England and France declared war on Germany. Dubos held up to the audience a bottle containing 500 grams of the antibiotic and dramatically announced that it was sufficient to protect five trillion mice against streptococcal blood poisoning. Florey rose to say that Dubos's first reports had encouraged the Oxford group to reconsider penicillin.

Waksman now seized the idea that his life-long career had prepared him for the task of experimentally seeking and exploiting antagonisms. He realized, he wrote later, that it was "relatively easy to isolate microbes which are able to kill disease-producing germs."[22] "I felt from my past experience that fungi and actinomycetes would provide far more effective antibacterial agents than the bacteria," he said. Having observed throughout his bacteriological studies that actinomycetes won out against other soil microbes in the struggle to survive under adverse conditions, he thought they might also destroy the germs in man, and he decided to center his investigations on the huge family of actinomycetes. "By a sheer stroke of good fortune," he would come to admit, "the actinomycetes proved to be highly productive of antibiotic substances."[23] Another motive for this research was the imminent World War II, which would require new agents to treat infections and epidemics. He had long ago refined techniques to isolate and culture soil organisms and to purify and crystallize soil-based substances. These served him well now.

The response of other researchers to gramicidin was scant. At a meeting of the Society of American Bacteriologists in late 1940, a proposed discussion Dubos organized on the subject of antibiotics, to be chaired by Waksman, was dropped because not enough participants wanted to attend. As 1940 slipped into 1941, and as the Battle of Britain raged on the other side of the Atlantic, Florey and Chain were feverishly pursuing the production of penicillin for clinical trials that proved its activity against gram-positive bacteria.

That same year, Waksman initiated a systematic and intensive screening program, unlike the chance discovery of penicillin by Fleming, for isolating, identifying, and testing potential antibiotics derived from soil actinomycetes, devoting much of the personnel and resources of his laboratory to the project. For financial assistance, he turned to private sources, particularly Merck and Company of New Jersey, a pharmaceutical firm for which Waksman had been a consultant for some time. Merck agreed to provide research and development assistance for any promising antibiotics in exchange for exclusive assignment of patent rights.

In his quest for a safe antibiotic that would destroy the germs against which penicillin was ineffective, Waksman was prepared to unearth and screen thousands upon thousands of soil microbes. To cultivate them on various media and test their ability to inhibit the growth of pathogenic bacteria and be safe for clinical use was a painstaking process.[24]

Waksman set his sights on finding an agent useful against gram-negative bacteria. Among the scourges falling into the gram-negative category are the bacteria that cause typhoid, bacillary dysentery, brucellosis, tularemia, and the bubonic plague. Against them penicillin is useless. Waksman was sure he

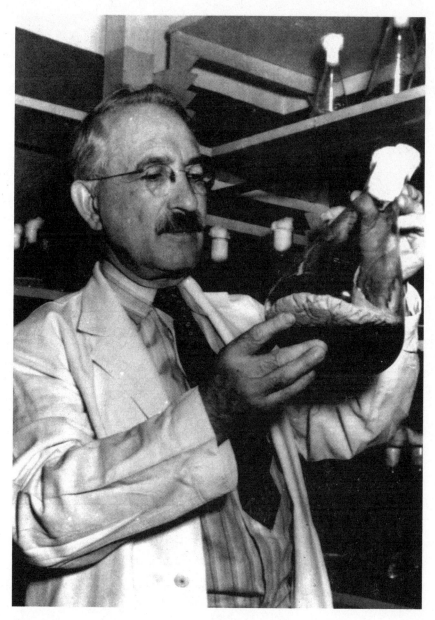

Figure 5.2   Waksman examining a culture flask for growth. [WA]

could unearth an antibiotic. "The stakes were high," he remembered in his autobiography. "And the game seemed worth the chance." Almost at once, in 1940, the research team isolated actinomycin and exulted in its accomplishment. But their hopes were dashed when testing in mice proved that it was

Figure 5.3 Boyd Woodruff, Waksman's research assistant, observes him following a step in the protocol established for searching and testing for an anti-bacterial substance. [WA]

highly toxic. Waksman was undaunted. Two years later, he discovered another anti-bacterial drug he called streptothricin, a promising agent against gram-negative microbes with almost incidental evidence of its efficacy against some mycobacteria—a family of bacteria that includes the one responsible for tuberculosis.[25] Excitement again coursed through the lab. But plans for human trials had to be abruptly cancelled when the animals they were experimenting on died. Yet Boyd Woodruff, Waksman's research assistant, helped to work out the methodology of how to search for and test the antibiotic. He was credited as co-author with Waksman in reports and a partner in patent applications.

Waksman was prospecting for gold and knew he was on the right trail.

The Scourge of Tuberculosis

One specter hovered over all the talk of gram-positive and gram-negative bacterial infections: tuberculosis, its fearsome microbe identified through a special acid-fast staining technique. Although penicillin worked wonders against many different types of bacteria, it proved useless against tuberculosis.

Tuberculosis is one of the oldest infectious diseases, having afflicted humans since Neolithic times. Since 1800, it has killed more than one billion people worldwide—more than every war, famine, and other epidemics put together. The "white plague of Europe" that raged in the seventeenth century was due to growing urban populations. Tuberculosis is an easily acquired infection. Coughing, sneezing, or even yelling or singing by an individual harboring the condition expels droplets containing tubercle bacilli into the air. Those can actually linger in the air for hours and be inhaled by others.

Mycobacterium tuberculosis was identified in 1882 by Robert Koch, a German whose painstaking work transformed bacteriology into a scientifically based medical discipline. The tiny rod-shaped organism was at that time responsible for one in seven of all European deaths. Its cell walls are neither gram-positive nor gram-negative, and an "acid-fast" staining technique allows ready microscopic identification. In 1905, Koch received the Nobel Prize for physiology or medicine for his groundbreaking work with bacteria. It was in the nineteenth century—when the Industrial Revolution shifted vast numbers of people from farms to crowded, filthy city slums—that tuberculosis surged. The international symbol for tuberculosis is "✝," symbolizing the many graves that it helped to create. The bacillus is singularly well adapted to survival. It is protected by a relatively impenetrable capsule. Because it has the ability to enclose itself in nodules, tubercles, in the body, the bacillus can remain dormant for years, then strike any part of the body with deadly results, although it most commonly affects the lungs.

For centuries, "consumption," so-called because the bodies of its victims slowly wasted away, was believed to physically stimulate intellectual and artistic genius. Some of the greats who were so afflicted include Molière, Voltaire, Spinoza, Goethe, Kafka, Chekhov, Paganini, the Brontë sisters, D. H. Lawrence, Thoreau, Poe, O'Neill, and Orwell. Characterizing the intellectual output of some consumptives was a certain melancholy or mournfulness—the music of Chopin and the poetry of Keats—in anticipation of a life to be cut short. Exemplifying the wide currency in the arts of the dramatic theme of the consumptive are the beautiful heroines Violetta in Verdi's *La Traviata* and Mimi in Puccini's *La Bohème.*

But the disease itself is far from romantic. It results in a relentless cough, bouts of fever, drenching sweats, typically at night, and utter exhaustion. A telltale sign is coughing up bright red oxygen-rich blood from an artery bleeding into the lung. In the condition known as scrofula, a word as harsh as the complication it describes, lymph nodes in the neck drain pus and cause ulceration of the overlying skin, and perhaps scar formation. An illustrious scrofula sufferer was Dr. Samuel Johnson. Tubercle bacilli may also spread to the intestines, kidneys, and bones (destruction of the spine with collapse and deformity is called Pott's disease). When the disease invades the brain, a critical condition results called tuberculous meningitis.

The pasteurization of milk, the establishment of sanatoria on wooded mountaintops or in deserts to provide fresh air and sunshine, and the posting of "No Spitting" signs were all public health measures taken to control the disease. In Thomas Mann's *The Magic Mountain*, the hero, Hans Castorp, is sent to a sanatorium high in the Alps. At the height of the movement in the 1930s, there were more than 600 sanatoria in the United States alone, with a total of 84,000 beds. The famed Saranac Lake Sanatorium in the Adirondack Mountains in upstate New York accommodated patients like Robert Louis Stevenson and the New York Giants pitching ace Christy Mathewson. But as to treatment, in the first decades of the twentieth century, physicians were limited to such homespun remedies as cod liver oil or aggressive methods of collapsing a lung. Patients dosed themselves with everything from pig's pepsin to iodine and copper. In the early 1940s, as the first reports on penicillin circulated, it became clear that the new wonder drug was ineffective against not only gram-negative microbes but tuberculosis.

But Waksman would find the cure in a place where no one had thought to look. He would find it in dirt.

In the past, he had stumbled across several opportunities to pursue the antagonistic effects of soil microbes upon the tubercle bacillus. But at the time he had neither the focus nor the resources. Now he set a determined course for unearthing antibiotics and could view these missed opportunities as stepping stones. The support from Merck in funding for his laboratory, skilled personnel, and capabilities for animal experiments and large-scale production provided a bedrock.

On June 1, 1943, he publicly committed himself to finding the cure for tuberculosis. At a meeting of scientists and representatives of pharmaceutical companies in New York to discuss the problems of treating the disease, a colleague suggested that a cure might be found in a certain enzyme derived from an earthworm. Waksman responded that the idea was absurd and was challenged by the colleague's angry reaction: "How do you propose to go about this problem?" Waksman's answer was as spontaneous as it was prophetic. "The antibiotics will do it," he said. "Just give us time."[26]

Little did he believe that the discovery would be forthcoming in only a few months.

There was, however, one major problem. Selman Waksman harbored a deep fear of contracting tuberculosis.

This explains why, when he was approached in 1932 with a grant from the National Research Council and the National Tuberculosis Association for $3,500 to investigate the bacillus's viability in soil, he assigned the task to one of his graduate students, Chester Rhines. The fact that Rhines was the sole author of three publications on the study in 1935 underscores the distance Waksman chose to maintain from the project. And, further, he chose to neglect to follow up on the unexpected finding that a soil fungus was noted to kill the tubercle bacillus. On a personal level, the death of René Dubos's wife in April 1942 from tuberculosis was a sad confirmation that Waksman's fears were not unfounded.

CHAPTER 6

THE STAR PUPIL

[In the laboratory] science is carried out largely by youth...from the age of 20 to early mid-thirties. They're the engine room of research activity.
—*Sir Paul Nurse, Nobel laureate in medicine,*
former president of Rockefeller University

Many bright and scientifically oriented people during the 1940s simply could not pursue post-graduate work because of either the expense or racial or ethnic quotas. Some saw in state-supported agricultural colleges an opportunity to obtain a chemist's education at low tuition rates with the likelihood of a career in academia or industry. This was especially true of those who grew up on farms and observed the problems of soil fertility and productivity. Albert Schatz came from this background. He was raised on a farm in Norwich, Connecticut, during the Depression, when his mother's refrain was, "Eat it up, wear it out, make it do, or do without."[1] After earning a bachelor's degree in soil chemistry at Rutgers, Schatz began post-graduate work with Waksman but was shortly drafted into the army. He served as a bacteriologist at a Florida military hospital where he grievously witnessed the ineffectiveness of penicillin in gram-negative infections. The most lethal of those were meningitis and typhoid fever.

Schatz's wife Vivian, who had graduate training in microbiology, recently told me about the reaction of her young husband to witnessing the devastation of infectious disease:

> At the military hospital, he saw the tragic deaths in young soldiers from
> gram-negative infections. The priest or minister would give last rites, but they
> wanted to talk to someone. Albert would sit by their bedsides for hours, listen-
> ing to their recollections and wishes.[2]

Figure 6.1   Albert Schatz's graduation photo as Bachelor of Science. [VS]

Obsessively, in his off-duty hours, he isolated and tested mold and actin-omycetes from contaminated blood culture plates and from Florida soils and swamps, and he sent these on to Waksman in his lab. Because of a back injury, Schatz was discharged after only a few months in mid-June 1943 and returned to Rutgers to complete his Ph.D. With Waksman's agreement, he persisted in looking for an antibiotic that was safe to use against gram-negative bacteria, continuing the program initially undertaken by Woodruff, who left Rutgers the year before to embark on a career at Merck.

To Waksman, the timing could not have been more auspicious. Waksman openly regarded Schatz as a "star" among his student researchers. He probably saw in the intense, fiercely motivated twenty-three-year-old many of the characteristics that had propelled himself to the pinnacle of his scientific field: initiative, resourcefulness, dedication, and passion. As with Waksman, Schatz had won a scholarship to attend Rutgers and had been elected to Phi Beta Kappa. On a personal level, Waksman could identify with Schatz's Russian-Jewish heritage and admired his avidity as a reader of the scientific literature, including the contributions of Russian scientists in the original language. As Waksman could recall singing revolutionary songs against the tsar as he crossed the Russian-German border a few years after the failed revolution of 1905,[3] so could he admire the socialist and humanitarian leanings Schatz evidenced.

Waksman himself had been greatly influenced in his early college years at Rutgers by one teacher in particular, a plant biologist by the name of Byron Halsted. They spent many hours discussing approaches to a scientific problem and the differences among universities and research organizations. As graduation approached, Waksman was elected to Phi Beta Kappa but could not afford to purchase the symbolic key.[4] Halsted gave him the fraternal grip and took from his pocket his own key, but being a Quaker, was restrained from displaying any decoration. He gave Waksman a five-dollar gold coin as a gift to be made into a key. "I was so overcome with emotion that I could only utter a few words in reply," Waksman said in his autobiography. He was so touched by the generosity of his mentor that he not only kept a framed photograph of Halsted on his office wall but also named his son after him.

For his part, Schatz, along with the other graduate students in the laboratory, had abiding respect for Waksman. "To us," he said, "he was a God-like figure, a person we almost worshiped."[5]

But the relationship between Waksman and Schatz proved in time to be complex, nuanced, and strangely co-dependent.

PAY DIRT

Research stipends then were quite meager. As an impoverished first-year graduate student, Schatz received $40 a month and survived by living rent-free in a small room in one of the plant physiology greenhouses in exchange for maintenance chores. He often relied on free fruit, vegetables, and dairy products from various departments of the Agricultural Experiment Station. Schatz chafed at the realization that he was being paid the least among all the graduate students. In truth, his unexpected abrupt return from the army caught Waksman short of funds, since he had used the budget line to hire another student, Elizabeth Bugie. Over the next several months, scraping money together from other sources, Waksman was able to raise Schatz's stipend to the customary $90 a month. Schatz describes himself at the time as skinny, weighing 120 pounds.

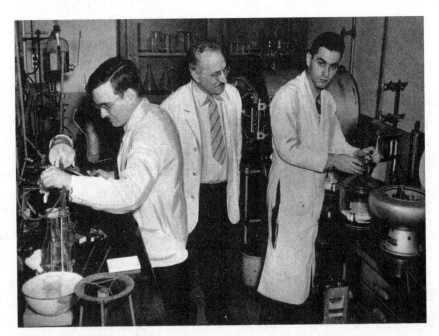

Figure 6.2 Selman Waksman overseeing research by Walton Geiger, the lab's chemist (left) and Albert Schatz (right). [WA]

When the decision was made to widen the screening program to include the search for an antibiotic against the organism responsible for tuberculosis, Waksman assigned Schatz to work in the basement laboratory, which was reasonably equipped. But he would be alone there, without an assistant or a colleague. According to Schatz, Waksman instructed him to never "bring a culture of the tubercle bacillus to the third floor," the site of his office and two other small labs. And Waksman never visited the basement laboratory."[6] Periodically, Schatz would report his activities to Waksman, who spent most of the summer on a working vacation at the Woods Hole Oceanographic Institution at Cape Cod in the study of marine bacteria.[7]

Schatz often had to tend his cultures overnight and slept in his laboratory. His descriptions of the conditions are Dickensian—sleeping on the floor with a torn blanket, at other times sleeping on a wooden bench only to waken occasionally to monitor his extractions. He prepared his own media and washed and sterilized the glassware he used.

Continuing the general research approach pioneered by Woodruff, Schatz analyzed hundreds of soil samples from a variety of sources. In a surprisingly short time, he found a microorganism that produced an antibiotic. He cultured a strain of the microorganism from a Petrie dish given to him by Doris Jones, his closest friend among his fellow graduate students. She was working with Rutgers's poultry pathologist and thought that a culture from a cotton swab taken from the throat of a healthy chicken might be of interest to him. Time would show that this instance would be distorted by Waksman to magnify his role in the discovery of streptomycin and to minimize Schatz's contribution.

Only three days later, Schatz found a second strain of the microorganism in a soil sample from a heavily manured field. In time, Waksman would manipulate this amazingly brief interval to cast doubt on the authenticity of the discovery. In a remarkable twist of fate, Waksman himself had found this same organism twenty-eight years earlier while he was working on his doctorate, but he had not pursued it. (The particular strain that he had isolated would be shown not to produce any antibiotic.) Because of the bacteria's greenish gray colonies he had called it *Actinomyces griseus*.[8] Since Waksman had recently changed the name of the species to *Streptomyces*, the organism would now be known as *Streptomyces griseus*.

Figure 6.3   Albert Schatz working alone in basement lab researching organisms against tuberculosis. [WA]

Tests quickly showed that the organism not only was active against gram-positive staphylococci but, most excitingly, also had dramatic killing effects on the gram-negatives. Clearly, the organism generated a powerful antibiotic. Waksman named it streptomycin. On Waksman's instruction, Doris Jones tested it in living tissue against gram-negative salmonella infections in chick embryos and confirmed its ability to cure. Furthermore, she found it had no toxicity in laboratory animals.[9]

Schatz's colleagues insisted that the heavy wax capsule of the TB bacilli would resist drug action. With an almost religious zeal to find a cure for tuberculosis, Schatz held to the premise that if nutrients and waste products could enter and leave the cell, so could antibiotics. He tested streptomycin against what appears to be a harmless avirulent strain of tubercle bacilli used in college courses and patiently waited several weeks for the slow-growing colonies to appear on the culture medium. On October 19, 1943, at about 2:00 in the afternoon, in the words of Doris Jones, "Despite all the odds, Al hit pay dirt!"[10] Schatz later recalled, "I realized I had a new antibiotic...I sealed the test tube by heating the open end and twisting the soft, hot glass...I felt elated and very tired."[11] This was just three and a half months after he began his search. The results were clear. Where streptomycin was added, not a single colony appeared, whereas in the controls dense growths were apparent. Schatz had shown that the new antibiotic was dramatically effective in inhibiting the growth of tuberculosis germs. As directed by Waksman, Elizabeth Bugie, the lab's lead technician, verified the results.

At this point, two determined researchers entered the scene, one from the Mayo Graduate School and the other from the Mayo Clinic. William Feldman, a widely recognized veterinary pathologist, notably in the fields of bovine and avian tuberculosis, and H. Corwin Hinshaw, a pulmonary physician, were aware of the dedicated search for antibiotics at Rutgers. Feldman was in a driven pursuit of a treatment for human tuberculosis and visited Waksman's laboratory in November 1943 to discuss the possibility that they might collaborate on studies involving potential anti-tubercular drugs.[12] He urged that a virulent strain of tubercle bacilli be included in Waksman's studies, rather than the harmless strains from the student bacteriology laboratory. Feldman soon sent a supply of highly infectious human tuberculosis bacteria to Waksman. Amazingly, Schatz's wife, Vivian, recalls that the material was simply sent as a parcel through the U.S. mail and was quickly passed on to Schatz for further testing.[13]

The discovery of streptomycin was announced in January 1944 in a short paper published in the *Proceedings of the Society of Experimental Biology and Medicine*, a rapid-publication journal, with the title of "Streptomycin, a substance exhibiting antibiotic activity against Gram-positive and Gram-negative bacteria."[14] The report is unusual in several respects. The names of the authors are Albert Schatz, Elizabeth Bugie, and Selman Waksman, in that order. No mention of tuberculosis is made in the title. The article spotlights the effect of the new antibiotic against gram-negative bacteria. The action of streptomycin upon twenty-two microorganisms receives attention but tuberculosis receives only marginal notation.

One medical writer, the respected physician Julius H. Comroe, Jr., was stunned by this: "The word 'streptomycin' is in the table, the words *M. tuberculosis* are also there, and the tenth line shows that streptomycin was an effective antibiotic against the tubercle bacillus. But nowhere else in the article (title, introduction, discussion, or summary) are the words *tubercle bacillus* or 'tuberculosis' mentioned again."[15]

The authors had clearly chosen an announcement as a soft whisper behind the hand rather than through a megaphone. Comroe flags this as "Discovery without Discovery." Some have presumed that Waksman simply didn't grasp the importance. Near the end of his life, Schatz offered a feeble disclaimer that he didn't want to raise false hopes since there were no data yet on toxicity or *in vivo* efficacy of the drug.[16] But the most likely explanation lies in the timing of Feldman's first visit to Waksman in November 1943.

Waksman's laboratory was now motivated to confirm streptomycin's activity against the virulent human strain. Schatz performed the studies based on test-tube work and, with Waksman as co-author, shortly published a second paper that concentrated solely on tuberculosis. Their conclusion was breathtaking: Streptomycin was fifty times more powerful than streptothricin in killing this deadly strain of human tuberculosis.[17]

SPECTACULAR RESULTS

Feldman was galvanized. Events moved quickly. Working day and night, Schatz used three stills twenty-four hours a day to concentrate enough streptomycin for initial animal testing. In a pilot study, Feldman used the precious sample of ten grams of the drug, supplied by Waksman from Schatz's feverish work, on four tubercular guinea pigs with very promising results.[18]

At this exciting development, Feldman and Hinshaw, recognizing the imperative of establishing priority and credit, displayed a remarkable act of professional courtesy. They chose to hold back the submission of their report of the guinea pig trials until Waksman could first publish his laboratory's work on the isolation of streptomycin. Both appeared in the *Proceedings of Staff Meetings of the Mayo Clinic*, Waksman's in the November 1944 issue,[19] based on an invited talk delivered at the Mayo Clinic a month before, followed a month later by the paper of Feldman and Hinshaw.[20]

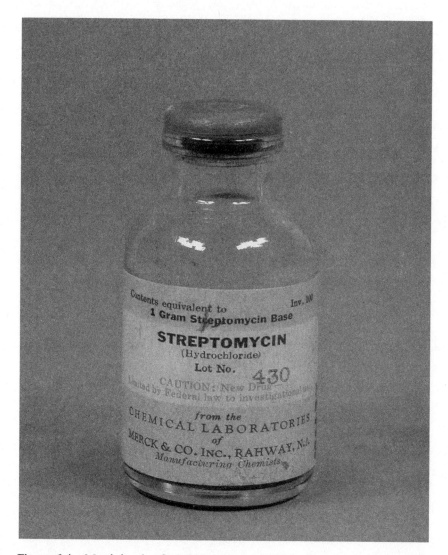

Figure 6.4 Merck bottle of streptomycin: "Limited by Federal law to investigational use." [SI]

The supply of streptomycin from Waksman's laboratory ran out in June 1944. In a meeting with Waksman and representatives of the Merck Company, Merck agreed to supply more streptomycin, which Feldman and Hinshaw used on a larger group of animals with virulent tuberculosis with spectacular results. In contrast to gram-positive or gram-negative infections, studies of the development of tuberculosis and its chemotherapy required months. The first patient successfully treated with streptomycin was a twenty-one-year-old woman at the Mayo Clinic in the last stages of tuberculosis. This was initiated fifteen months after *Streptomyces griseus* was first isolated in Waksman's laboratory. The period from initial discovery of streptomycin to clinical application—the transitional time in medical parlance—was remarkably short.

Eight pharmaceutical companies provided an estimated $1 million worth of streptomycin for the largest clinical study of a drug ever undertaken, involving forty-six Veterans Administration hospitals, hundreds of investigators, and many thousands of tuberculosis patients. This constituted the first privately financed, nationally coordinated clinical drug evaluation in history. Medicine had taken a giant step in embracing the value of controlled experiments repeated thousands of times in the evaluation of a new drug. Once again, the drug worked wonders. Side effects, including an impaired sense of balance and deafness, proved to be transitory and could be minimized by controlling the dosage. It was also effective against bubonic plague, cholera, typhoid fever, and other infectious diseases caused by gram-negative bacteria. In 1947, streptomycin was released to the public.

Streptomycin buried the nihilistic concept that *nothing* would ever kill tubercle bacilli in humans. But, in time, it would become evident that unfortunate side effects, mainly a potential loss of hearing, could develop and that the tubercle bacilli could develop a resistance to streptomycin. Two other drugs entered the picture—para-aminosalicylic acid (PAS), developed in the mid-1940s by Jorgen Lehmann, a pharmacologist and cell biologist in Sweden, and isoniazid (INH), developed later by the pharmaceutical firm of E.R. Squibb and Sons, which interfered with essential bacterial biochemical processes. One grateful recipient of combined therapy was Feldman, who contracted pulmonary tuberculosis in 1948. Hinshaw, his close friend and fellow scientist, treated him with streptomycin and PAS, and Feldman spent some time at a sanatorium in New Mexico. Feldman made a complete recovery. In December 1954, the world-famous Trudeau Sanatorium in Saranac Lake shut down its patient-care facilities and directed funds to its research library. The reason: not enough patients, a direct result of the success of drugs.

A few statistics illustrate how dramatic the effects of anti-tuberculosis drugs were. In 1944, before streptomycin was introduced, there were 130,000 new cases of tuberculosis in the United States. By 1989, there were fewer than 24,000 cases.[21]

Waksman was hailed as a medical hero—the discoverer of the world's newest "miracle drug"—whose victory over tuberculosis resonated with symbolic value in the wake of World War II. In the resulting avalanche of public acclaim, he toured the world, gave lectures, and took tours of medical facilities. Although Schatz, the graduate student, had actually made the discovery, it was Waksman who had designed and supervised the screening program and, as head of the department, was in a position to arrange for commercial development. It was exactly at this juncture that trouble began.

THE RUPTURE

The success of streptomycin also brought problems in its wake: Who would reap the rewards of its discovery? In January 1945, Rutgers filed a patent application for streptomycin in the names of Selman Waksman and Albert Schatz. Two months later, as Schatz was preparing to defend his Ph.D. thesis on streptomycin,[22] the young graduate student married and took his test tube cultures, covered with cotton plugs, along on his honeymoon so he could continue his observations. He stayed on at Rutgers for a year after his Ph.D., working on an antiviral research project while his wife completed her senior year at the New Jersey College for Women at Rutgers.

The press conducted multiple interviews with Waksman, who never indicated Schatz's role. Schatz began to chafe under the perception that he was being slowly but inexorably marginalized. Waksman continued to refer to "my assistants and I" but never acknowledged Schatz by name.

Figure 6.5   Selman Waksman (right) conferring with Albert Schatz. Following the discovery of streptomycin, this is the last photo of them together before Schatz's role was marginalized. [WA]

Schatz felt impelled to intercede and resorted to a ruse. Posing as his uncle, Dr. J. J. Martin, a dentist, Schatz wrote to the editors of several journals and medical newsletters, admonishing them for not including the role of Albert Schatz in descriptions of the discovery of streptomycin. But the attempt to assert his contribution failed miserably. Informed of these letters by his colleagues, Waksman called Schatz into his office and demanded an explanation. In Waksman's account, Schatz became shamefaced. He "burst into tears" and said that his uncle was dedicated to making his nephew famous. Schatz offered to ask the journals to retract such letters.[23]

Schatz's entire career depended on good recommendations from Waksman, and both men knew it. At the Waksman Archive at Rutgers, I found a farewell letter from Schatz dated May 21, 1946, immediately after the above episode. It lays no claims to priority and includes the key statement:

> The work on streptomycin, carried out under your guidance and continuous active participation . . . will stand as a symbol to cooperative work under wise and able leadership.[24]

Rather than merely an expression of gratitude, the letter firmly endorses Waksman's preeminent role. Schatz would later claim that he was bullied into this or else Waksman would "see to it" that he would never find employment. He claimed that Waksman dictated the letter in both senses of the word.

For his part, Waksman was not above using stratagems. To reduce Schatz's claims and establish himself as an essential participant in the steps immediately leading to the discovery of streptomycin, he fostered two myths:

(1) In truth, Schatz had directly received the throat culture taken from a healthy chicken by his fellow graduate student. Later, Schatz postulated that "the chicken had probably inhaled a spore of *S. griseus.*"[25] Waksman, however, created a fanciful account that the culture had been taken from a wheezing "sick chicken" whose throat was partially clogged with a growth of *Streptomyces griseus.* He further claimed that the poultry pathologist gave the culture directly to him, and he then passed it on to Schatz. Waksman had a proprietary interest in this particular species since he had identified it in 1915. (A similar attachment was displayed by Alexander Fleming with penicillin. Hearing of the Oxford's team investigation of its therapeutic potential after eleven years of its virtual neglect, Fleming visited Florey and Chain and inquired, "What have you been doing with my old penicillin?")[26]

Waksman first outlined this fable in May 1946 in a letter to the editor of the international *Medical Digest.*[27] It has been recounted over the years in numerous magazine articles and books until it hardened to something firm enough to pass for truth.[28]

(2) Schatz shortly discovered another strain of *S. griseus* in a sample of heavily manured soil. This also yielded streptomycin of even greater potency and was used subsequently for animal tests. Waksman, however, openly speculated that floating spores from the first culture had contaminated Schatz's soil sample to grow the microbe.[29]

These myths ("fairy tales" in Schatz's estimation) were repeated so often that Waksman convinced himself of their truth.

Schatz left Rutgers in May 1946 and over the next couple of years grew frustrated and distraught at his career. He maintained a correspondence with Waksman, informing him of his research activities and on occasion sending him cultures he thought would be of interest. Waksman sent monographs and reprints of articles and offered letters of recommendation. Schatz set his sights on working as a postdoctoral fellow with a well-known researcher at the

Hopkins Marine Station in Pacific Grove, California, an affiliate of Stanford University. Again relying on Waksman's recommendation, he wrote:

> I have not the slightest desire for fame, glory, popular acclamation, or a lot of money. I want to do the work I like and feel good about it "inside of me"... my pocket and stomach have been full but my head and heart have become empty.[30]

Schatz worked there in late 1948 and early 1949. The GI Bill added a marriage bonus of $90/month to his meager income while he and his wife, Vivian, lived in spartan conditions with no hot water. Photographs show that Schatz adjusted, growing a bushy beard to avoid the need for shaving. Pointing out that "our savings account is at $800.00," he explained to Waksman:

> Those things we value most cannot be purchased with money. To be perfectly honest with you, I simply would not know what to do with more money if I had it, and I sincerely hope that I shall never have to spend any time attempting to solve such a problem for myself... It is impossible for me to over-evaluate what your friendship, confidence, and encouragement have meant to me. I am so constituted that those things are worth far more than material aspects.[31]

Little did he suspect that these words would come back to haunt him.

In a seemingly innocuous request, Waksman sent Schatz a document assigning his U.S. patent rights for streptomycin to the Rutgers Research and Endowment Fund. Without any suspicion and assuming that Waksman was also freely donating his rights to Rutgers and that no one was profiting from streptomycin, Schatz signed the authorization.

For his part, Waksman had given the rewards of streptomycin considerable thought. Rutgers was able to maintain the patent and the royalties from the drug's sales, based on nonexclusive licensing, yielding a huge financial windfall for the university and Waksman personally. His vision was to establish a well-funded self-sustaining institute of microbiology, a center for microbiological research and graduate instruction.

MONEY MATTERS

Two events shocked Schatz, even as he became increasingly aware that Waksman was garnering praise and recognition in magazines as the discoverer of streptomycin as his own role was being eclipsed. In November 1948, Waksman won the highly prestigious Albert Lasker Medical Research Award, which often presages the Nobel Prize.

Earlier in the year, Waksman sent Schatz a check for $500. Schatz thanked him and offered to repay him. Over the course of 1948, two more checks followed, of $500 each, with the understanding that Waksman was generously sharing some small funds he received from the Rutgers Research and Endowment Fund for the discovery of streptomycin. Waksman never mentioned the royalties he was receiving. These were, however, personal checks, and it struck Schatz that they should have come directly from the institution itself. He filed an income tax return based on the impression that the money was a gift to supplement his income while a postdoctoral worker in California and was therefore not taxable. The tax status came into further question when Schatz received a 1099 form from Waksman himself for the amount under the heading "salaries, fees and commissions." Schatz, however, had never been employed by Waksman and had always been paid by Rutgers. Afterward, Schatz would come to believe that Waksman wrote the checks because he felt guilty. "He keeps me as an illegitimate child whom his conscience compels him to support secretly."[32]

The second clue came on January 12, 1949, when in a brief business-like letter Waksman requested that Schatz sign over to him foreign patent rights. The form for release of Canadian rights was accompanied as a legal device by a notation for $1 and the release for New Zealand for $100, "receipt of which is acknowledged hereby," even though no form of money was included with the papers. Waksman offered no explanation for the assignment.[33] He added, incidentally, that he was continuing to submit Schatz's name to openings in bacteriology at various schools of medicine.

A more heavy-handed approach by the university administration and Waksman cannot be imagined. In fact, the strategic course the Rutgers Research and Endowment Fund and Waksman had chosen had the unintended effect of increasing the breadth and complexity of the issue.

Ten days later, after much agonized deliberation with himself, Schatz sent Waksman a seven-page, single-spaced typed letter.[34] The tone is initially one of deference, solicitude, and cautious restraint, but the letter clearly indicates a dawning sense of unease and suspicion. He reminds Waksman that at the time they signed the original patent application for streptomycin four years before, "I knew absolutely nothing about such matters and was relying entirely upon your judgment [with] complete confidence in you, and for this reason I have unhesitatingly and unquestioningly signed any and all documents on which you requested my signature." But a disturbing realization had grown: "I have slowly been signing myself out of streptomycin completely." He came to see the Rutgers Research and Endowment Foundation as a shadowy channel about which he knew virtually nothing and which had not responded to any of his recent direct inquiries. He asked Waksman for elucidation. What is it? What is its purpose? Who are the officers and what are their salaries? Has the streptomycin patent yet been granted in the United States? (He had never been informed that the patent, U.S. 2,449,866, titled "Streptomycin and Process of Preparation," was issued to Schatz and Waksman four months earlier.[35]) What royalties has streptomycin generated? Has anyone personally benefited directly from these funds? What is the status of patent applications in foreign countries?

The scales had clearly been lifted from his eyes. A sense of embittered righteousness in two associates who had had respect and some affection for each other was building to an inevitable clash.

Schatz went on to ask for copies of all documents that bore his signature. He reminded Waksman that after signing the original patent application, they shook hands in a gentleman's agreement that as partners in streptomycin, "neither of us would profit financially from this discovery." He went on to say that his motivation in pursuing more information had never been materialistic, but rather was based on the esteem of his professional peers, observing that Waksman alone had been receiving scientific awards and honors for streptomycin. "I have always been concerned with my own personal evaluation of myself," but "torn and tormented" as these suspicions have grown, he now sought "peace of mind and self-respect" in order to "retain dignity in my own eyes and continue to live in harmony with myself."

Schatz ended the letter with the bold statement that "regardless of what will transpire, I can promise you that I shall be true to my own convictions."

Yet, after all of this, in a handwritten P.S., as if in a child's attempt to placate a father, he added brief comments and personal notes to update Waksman on mutual acquaintances and previous graduate students, and he indicated his intention to send Waksman reprints of one of his recently published papers.

Waksman was just as pointed in his lengthy reply, in a tone Schatz must have found arrogant and condescending. After expressing pain and outrage over Schatz's brashness, "You were one of the many cogs in a great wheel," he said sternly, "my tools, my hands."

He reminded Schatz that the student in his laboratory "isolates these cultures as a part of his process of training" and that "all the methods for the isolation of streptomycin, namely those used in the isolation of streptothricin, have developed in our laboratory long before your return from the Army...the stage was all set for a startling discovery...In this, you have made only a limited contribution, but you have reaped far more glory than you were entitled to...[your] assisting in the development of the methods for the isolation of the crude material from the media and in testing its antibacterial properties...was only a very small part of the development of streptomycin."

He continued, "I was as generous as any Professor could be expected to be...I hardly needed to worry for my own share of the credit. After all, what greater glory can come to a teacher than that his students have done well and have carried on the torch along the path that he has blazed...The fact that your name was placed first [on the streptomycin article]...was largely a courtesy to a hardworking and diligent assistant."[36]

Waksman made the point that "others," namely Boyd Woodruff, had earlier assigned rights in the patents for actinomycin and streptothricin to the Rutgers Research and Endowment Fund. He had always considered this as acceptable protocol. U.S. patent law requires that a patent application be executed by all co-inventors. Thus, the co-authors of a scientific publication first announcing a discovery must co-sign the patent application detailing the steps leading to the discovery. Here, there's an important distinction. Streptomycin, a natural substance, cannot be patented, but the steps necessary in its isolation and preparation can. It was always expected that the assistant execute an assignment of his rights, if any, in the invention.[37]

"How dare you now present yourself as so innocent of what transpired when you know full well that you had nothing whatever to do with the practical development of streptomycin and were not entitled to any special consideration." Waksman concluded the six-page letter by urging Schatz to reconsider. "You have a great future before you, and you cannot afford to ruin it."[38] Paternalistic advice or a veiled threat?

The very next day, in an "Oh, by the way" note, as if seeking Schatz's understanding, Waksman informed him, "I failed to mention that the funds collected by the RREF are to be largely used for the building of a great Microbiological Institute."[39]

Schatz also shortly received further elucidation from Russell Watson, attorney of the Rutgers Research and Endowment Fund. He explained that the fund was a non-profit corporation whose net income was devoted solely to the promotion of scientific research at Rutgers University and that the trustees of the foundation are also trustees of the university. He did not mention that he himself was a trustee. Watson added that the U.S. patent for streptomycin, dated September 21, 1948, was issued to the foundation and that foreign applications were pending in Canada, New Zealand, and Argentina.[40]

This made it clear to Schatz that Waksman and Rutgers had moved quickly and stealthily.

He now had no doubt that he was quickly sinking into scientific obscurity while the professor collected prizes and honors. Rutgers's plan to establish the Institute of Microbiology was officially announced to the press on May 5, 1949. In Schatz's mind, as noble as the goal was, it was being made possible by the bounty of *his* discovery, unacknowledged and unappreciated. He was driven to devise another ruse and enlisted a family friend as an accomplice.[41] In June 1949, many of the previous graduate students who worked in Waksman's laboratory at Rutgers received a letter from MD Bromberg International Publishers with offices in New York City. The letters stated that the company would be publishing a book detailing the discovery of streptomycin by Dr. Albert Schatz and solicited comments, recollections,

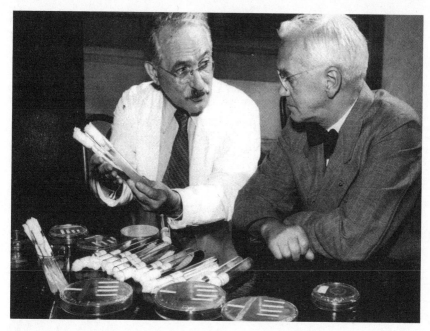

Figure 6.6   Alexander Fleming, the discoverer of penicillin, visits Selman Waksman for a private tutorial on streptomycin. [WA]

and personal endorsements of his major role.[42] Most of the recipients saw through the charade, several communicated with Waksman, and the Schatz Archive contains only one positive response—from Schatz's loyal friend, Doris Jones.[43]

Around this time, Nobel laureate Alexander Fleming visited Waksman to observe the new miracle drug. Waksman was featured on the cover of the November 7, 1949, issue of *Time* magazine. There was no doubt that this was the opening shot in his international lionization. Schatz's sense of injustice was inflamed.

UNPRECEDENTED ACTION

Then something happened that thoroughly rocked the scientific community. On March 10, 1950, Albert Schatz filed a legal claim in New Jersey Superior Court demanding formal recognition as co-discoverer of streptomycin and a share of the royalties. This widely publicized lawsuit by a former doctoral student against a distinguished, internationally recognized professor was unprecedented. Even a casual browser of the *New York Times* would be stopped short on encountering the headline, "Streptomycin profit asked by ex-student."[44] For Waksman, the lawsuit came as a shocking, vicious, and humiliating blow.

Schatz alleged that Waksman told him to sign over his share of the streptomycin patent to the foundation or he would "see to it" that the graduate student never got a job by virtue of his power, position, and influence in his field. Schatz also declared that Waksman assured him no one would profit personally from the drug royalties. As reported in national newspapers, this cast a grave shadow on Waksman's integrity.

In a letter of support, Robert C. Clothier, president of Rutgers University, labeled Schatz's action as "a most contemptible piece of ingratitude" and ended with a note of reassurance to Waksman: "The matter is in the best possible legal hands and...can have no effect upon our ultimate purposes."[45] This phrase clearly indicates that much planning had been undertaken behind the scenes. Indeed, the bare facts revealed point by point during the pre-trial deposition process shocked Schatz.

Five years earlier, in February 1945 (the same month the Waksman and Schatz patent application for streptomycin was filed), Waksman, along with William Martin, the director of the New Jersey Agricultural Experimental Station, persuaded Merck and Company to rescind its exclusive contract with Rutgers University, dating back to November 1940, regarding the development and marketing of antibiotics.[46] Such exclusive arrangements between academic scientists and drug companies were by no means unusual by this time. Under this licensing agreement, Merck supplied chemical assistance, animal testing facilities, and large-scale production apparatus. Fleming's penicillin had languished for over eleven years as a laboratory curiosity before its widespread use in human infections was made possible. Waksman's relationship with Merck guaranteed this would not happen to him. But Waksman was now concerned that the benefit of life-saving properties of streptomycin against gram-negative bacteria and tuberculosis, particularly with wartime needs, could be accomplished only from more widespread production.

Just two years earlier, Merck heavily committed—along with Pfizer, which took on the dominant role, and other pharmaceutical companies—to producing quantities of penicillin for wartime needs. George Merck had undertaken this participation, putting public spirit before profit. In a telegram on July 26, 1941, to Alfred Newton Richards, national chairman of the committee seeking help in the wartime production of penicillin, Merck stated, "Command me and my associates... if you think we can help you."[47] Furthermore, Rutgers had recently become the State University of New Jersey, and huge exclusive commercial profits flowing into a public institution on the basis of monopolization would be deemed, it was thought, unseemly.

An imposing figure of six feet five inches, George Merck IV had taken over the family drug company at age thirty-two and quickly transformed it into a scientific powerhouse. He fostered research and development by allowing scientists to publish their work, whereas other pharmaceutical companies kept findings secret. George Merck was magnanimous, requiring only that the company withhold about $500,000 from future royalties for its research and development expenses for streptomycin. This exceptional gesture enhanced the Merck company's reputation, not only for its science but also for its humanitarian concern. It enabled contracts with additional drug companies intending to make streptomycin with the same royalty basis as Merck of 2½ percent. By May 1949, steadily rising production by eight American pharmaceutical companies, with Merck remaining preeminent, generated almost $200,000 a month income for the foundation. Royalties the previous year were more than $700,000.[48] Within a few years, streptomycin would also be manufactured by three companies in France, two in England, four in Japan, one in Sweden, and two in Italy, and several other plants were being erected in other countries.[49]

The Rutgers Research and Endowment Fund was organized in May 1946 and took over Waksman's antibiotic patents not only for actinomycin and streptothricin but also streptomycin. As soon as the latter's patent was granted, it was immediately assigned to the foundation. Waksman was to receive 20 percent of the royalties received by the foundation, but when sales of streptomycin skyrocketed, that amount was reduced to 10 percent. Schatz simply was never informed of these arrangements. By the time of the lawsuit, the foundation had received $2,600,000 from the royalties. But the stunning admission was that Waksman had received a total of $350,000 as "commission" from the foundation.

The choice of the word "commission" was unfortunate. It was intended as a form of compensation for Waksman's efforts, before the foundation was established, in personally taking on all the stressful demands of organizing the legal negotiations and license arrangements with the various companies as well as the practical developments of the production and utilization of streptomycin.[50]

Undoubtedly adding to Waksman's smoldering outrage was the realization that in none of Schatz's numerous letters in the two and a half years after leaving Rutgers did he mention streptomycin at all, nor did he show the slightest interest in its toxicity studies, the rapid progress in its large-scale production, the elucidation of its chemical nature, or its utilization in clinical medicine. During that time, Merck had assigned fifty of its scientists to the task and completed a large program of clinical trials from pilot plant production as well as provided samples to Feldman and Hinshaw for their work on tuberculosis at the Mayo Clinic. The letters from Schatz, and their lack of interest in these processes, reflect his self-absorption.

Rutgers Research and Endowment Fund attorney Russell Watson tried to put the $350,000 in a more favorable light: After paying $180,000 income tax on the amount and also donating $30,000 of it to Rutgers for research grants, Waksman, who had dedicated eleven years to working on antibiotics, derived an average annual take home pay of only $12,000-$13,000 from streptomycin.[51] By this time, Waksman's annual salary as a faculty member was $10,000. But the attempt was like using a toy water gun to extinguish a raging fire.

With streptomycin, Waksman won for his university not only fame but fortune. But the trajectory of his career was now being sidetracked from glorification to vilification.

To Waksman's great dismay, the fact that he received royalties from the streptomycin patents was leaked to the newspapers, and the case received wide publicity. A headline in the *Newark Star-Ledger* screamed, "Attorney says Waksman kept $350,000 streptomycin profit."[52] Particularly damning was the *New York Times* report of Schatz's charge that Waksman had given an impression that he turned over all return from the discovery to the university research foundation.[53]

Rutgers planned to build an Institute of Microbiology with drug royalties. Waksman felt that the "age of antibiotics" was only beginning, and he wanted to do what he could to speed its progress. He planned to donate $250,000 for the institute's operating expenses and would be the institute's director. What the Pasteur Institute did for bacteriology, Waksman hoped his institute would do for microbiology.

Regional newspapers were more cynical. *The Passaic* (New Jersey) *Herald News* was merciless in its condemnation of Rutgers. In an editorial entitled "Rutgers Is Too Smart for Its Own Good," it berates the university for allowing "the public to assume [Waksman] wasn't getting a nickel from the proceeds...Rutgers has been too clever for its own good...Rutgers must get over the idea it can hoodwink the public."[54]

PRE-TRIAL MANEUVERS

Waksman's testimony throughout the detailed deposition was vigorous and emphatic. He likened his direction of the work in his laboratory to that of an orchestra conductor, with a host of student assistants over the years, and repeatedly stressed that Schatz's role, while dedicated and hardworking, was neither creative nor singular. "Is the one, Dr. Schatz, who was instructed to isolate the cultures and test them, is his work more important than the man who washes the dishes the cultures were made in, to the girl who prepares the media...to the one who tests the material in animals, to the girl who grows the culture in a flask?" Hearing this, Schatz had the sardonic thought: "So what. One might also mention the carpenters, bricklayers and plumbers who built the laboratory."[55]

Nevertheless, Schatz's lawyer asked Waksman to estimate how much each lab worker had contributed to the ultimate discovery of streptomycin. While reluctant to do this, he did factor the worth of each of the many contributions to the project, perhaps more to indicate the relatively small role of Schatz. "If I would say my contribution would be roughly, let's say, 75 percent, Schatz's contribution perhaps 10 percent, Betty Bugie's contribution perhaps 5 percent, there was Doris Jones who first isolated the culture that produced streptomycin from the chicken's throat." Waksman's anger came to the surface. "Now, how much is the poultry pathologist entitled to for his contribution? How much is the chicken entitled to for its contribution, for having that culture picked from its throat?"[56]

Robert Starkey, professor of microbiology, served over many years as Waksman's trusted and loyal associate in overseeing the operations of the research laboratory. Along with Waksman, he had sat on the committee a few years earlier that approved Schatz's thesis on streptomycin and awarded him the Ph.D. In a letter to Waksman, then traveling in Europe, he expressed misgivings over Waksman's handling of the affair with Schatz and emphasizes that they are shared by many. It is an indication of the respect that Waksman commanded that Starkey, after twenty-three years as his full-time associate, still addresses him as "Dear Dr. Waksman."

> I am sure that your friends have no feeling that you are not justified in having any amount of money from the patents that you might wish but they feel hurt that they have been misled with regard to disposition of the funds... [Many] are apt to consider the matter as reflecting upon the integrity of persons involved.[57]

Displaying a remarkable sangfroid, Waksman left during the ongoing pre-trial hearing for a four-month trip to Europe, where he garnered numerous awards and honors. These included a silver medal from Pope Pius XII at an audience with the pontiff, a silver medal from the University of Florence, and an honorary doctor of science degree from the University of Madrid at a ceremony before 10,000 persons. He attended meetings of the World Health Organization in Geneva and the Seventh International Botanical Congress in Stockholm. In France, he spoke before the University of Toulouse and the National Tuberculosis Association of France. Most heartfelt were the numerous personal tributes by children and adults restored to health through the use of streptomycin.[58] Waksman's response was one of solemn pride. He returned to Rutgers in mid-August.

Waksman felt that the discovery of streptomycin was the last inevitable step in a path he had personally paved with the prior discoveries of actinomycin and streptothricin, and that Schatz, "a carefully supervised laboratory assistant," was lucky to be "in at the finish." It was Waksman, after all, who had established the entire program of antibiotic research, nurtured streptomycin through development, and oversaw arrangements for its marketing.

Depleted and embarrassed, Waksman agreed to an out-of-court settlement. The Rutgers Research and Endowment Fund wanted to press on, but "the situation made a speedy disposition of the lawsuit essential," Waksman later explained. "The decision to settle the suit was largely mine."[59] This decision on December 29, 1950, cut off the scandal and cut Rutgers's losses, whereas a trial by jury might have given Schatz unbounded compensation. Most of all, according to Vivian Schatz, the foundation feared loss of patent rights.[60]

Most importantly to Waksman and Rutgers, the settlement guaranteed the means to establish the Institute of Microbiology and Waksman's foundation. Ever practical, Waksman swallowed a bitter pill and acknowledged in court that Schatz was "entitled to credit legally and scientifically as co-discoverer of streptomycin." Schatz received an immediate lump sum of $125,000 for foreign rights, a public statement by Rutgers officially acknowledging him as co-discoverer of streptomycin, and 3 percent of the ensuing royalties. The latter provided roughly $150,000 a year for several years. At the time, the thirty-year-old Schatz was an assistant professor of bacteriology at Brooklyn College, with a salary of $4,980 a year. Waksman shared 7 percent of his 17 percent royalty with the other laboratory workers at Rutgers and his colleagues at Mayo. He took pains to include twelve laboratory assistants, clerks, and even the man who washed out the laboratory glassware. All in all, more than two dozen people received some compensation.

While many may see such a generalized distribution of funds as an act of extreme generosity, to Schatz it was simply a flagrant effort to trivialize his role in the discovery of streptomycin. Carried to its extreme as a legal maneuver, the allotments clouded the designation of anyone's central contribution other than Waksman himself. Two of those designated declined to accept their allotment. William Feldman cited the long-standing tradition at the Mayo Clinic not to accept money from outside sources.[61] And Walton Geiger, Waksman's chemist who had been on the staff for only a few months before the discovery of streptomycin, declined on the basis that scientific discoveries are for the public good and not for personal gain.[62] With half of the rest of the money, Waksman started the philanthropic Foundation for Microbiology, thus reducing his own portion to 5 percent. The settlement was broadcast on page one of the *New York Times*[63] and in *Time* magazine.[64]

In a face-saving statement, Clothier, the president of Rutgers, announced, "It has never been disputed that Dr. Schatz was a co-discoverer of streptomycin."[65] Schatz, like a victorious gladiator gloating over his fallen opponent, issued a statement to the press that in agreeing to the settlement, he had been "influenced by the fact that the [Rutgers Research and Endowment Fund] and the public trust imposed upon it might be jeopardized if the litigation were to continue to its ultimate conclusion."[66]

PYRRHIC VICTORY

Schatz felt vindicated and hoped to use the money to continue in microbiology research. It was, however, a pyrrhic victory. Following the notoriety, the doors of top-grade laboratories were closed to him. He was viewed as a malicious malcontent who had unjustifiably attacked a world-famous scientist. No other department head wanted to hire an aggrieved "whistle-blower," as such a contentious, publicly aired episode had been unknown within the scientific community. While Waksman was hailed worldwide as a medical hero, Schatz faced a firestorm of scorn. The establishment closed ranks. He was, in effect, blackballed.

History might view Schatz as naive and reckless, heedless of the possible consequences of his actions. But it was clearly an abandon fueled by a burning sense of injustice that he was a victim of guile and was not given appropriate credit for his contribution. Even as a student in Waksman's laboratory, he wanted intensely to be recognized. It was the only reward he craved. With great pride, he had brought specimens home to his family to demonstrate the nature of his research work. And afterward, he presented to his mother the sealed test tube of that fateful day in October 1943 that marked the discovery of streptomycin. In time, it made its way to the collection in the Smithsonian Institution. Of the more than fifty universities and research institutions he applied to for work, Schatz received only three offers, but over a weekend all three were withdrawn.[67] Finally, he found a position at a small private agricultural college in Doylestown, Pennsylvania, where he remained for several years. Its name, the National Agricultural College, was more grandiose than its academic standing.

Tucked away in one of the numerous boxes in the Waksman Archives at Rutgers, I found an intriguing citation by Waksman of a commentary on the

Talmud: *The student who insults his teacher is to be banished (excommunicated)*.[68] Whether this undated memo represents righteous declaration or, after the fact, certitude of his prediction, there is no doubt that Waksman found it unnecessary to foster "banishment" from the scientific community. It was an already de facto occurrence.

Few of the scientists who sided against Schatz knew the details of the case, however. Vivian Schatz knows in her heart that her husband was not primarily motivated by money. She told me, "Albert believed his work was for the common good and he would have never instituted the lawsuit if he didn't feel there was duplicity on Waksman's part. He did, though, want recognition for his discovery."[69] Years later, Boyd Woodruff would comment:

> At the time, there were few examples of how to handle the monetary benefits resulting from a major research breakthrough. Errors in judgment had been made in the approaches used at the university.[70]

But Waksman always had a very clear vision of how best to use the rewards of his work in establishing the Foundation for Microbiology. Its purpose: to support professorships and fellowships, finance the publication of technical papers and books, sponsor scientific meetings, and help in the distribution of microbiological discoveries for practical use. The foundation is described as "both a family and professional heritage." Selman Waksman served as its first president from 1951 to 1969 and was succeeded (1970 to 2000) by his son Byron Halsted Waksman, now a respected immunologist. The founder's grandchildren likewise serve on the foundation's board of trustees, along with leaders of American microbiology.

In France, Waksman convinced the manufacturers of streptomycin to contribute to a fund that was used to support microbiology in France. The same thing was done on a larger scale in Japan, where half of the Japanese royalties were assigned to the Waksman Foundation of Japan.[71]

The knockout blow for Schatz came in October 1952, with the announcement that Waksman would be the sole recipient of the Nobel Prize in physiology or medicine. Although the prize can be given to three people, Schatz, along with other possible candidates such as René Dubos or Feldman and Hinshaw, was conspicuously passed over.

The final wording of the award to Waksman was carefully chosen: for his "ingenious, systematic and successful studies of the soil microbes that led to the discovery of streptomycin," instead of "for the discovery of streptomycin," as had first been announced. Burton Feldman, a historian of Nobel awards, uncharitably refers to Waksman as getting the Nobel "for not discovering streptomycin."[72]

Schatz could not rid himself of a festering indignation and waged an unsuccessful campaign to obtain a share of the prize. He circulated requests for support to many, including Nobel laureates Howard Florey, Hans Krebs, and Otto Loewi. Albert Sabin of polio immunization fame responded directly:

> It seems to me that Dr. Schatz should have considered himself to be an unu-
> sually fortunate graduate student in having been permitted by Dr. Waksman
> to participate in the great work he was doing... In my opinion, Dr. Schatz is
> behaving like an ungrateful, spoiled, immature child. When he grows older,
> he will regret what he has done and is doing now.[73]

The Nobel committee tactfully answered a complaint from the vice president of the National Agricultural College:

> It was generally regretted that part of the information in your letter had
> not been accessible to the members of the faculty, since it had not been
> published in any scientific journal. It may be of interest to you that numer-
> ous American colleagues... have suggested the name of Doctor Waksman,
> though none of them has proposed Doctor Schatz.[74]

In an astonishingly audacious effort, Schatz also petitioned King Gustav VI of Sweden.[75] At the Nobel ceremony in December, Waksman was presented the award by the king with the words, "We regard you as one of the greatest benefactors to mankind."

The public dispute was tragic for both men. Waksman, heartbroken, regarded the year 1950 as "the darkest one in my whole life" because of the perceived betrayal by Schatz and the horrible publicity brought about by the whole affair.[76] Schatz, a penniless student, was at least financially recompensed but was deprived of what he felt to be his rightful share of the Nobel Prize and of career possibilities.

By 1954, royalties from streptomycin amounted to about $5 million. Waksman's second and major goal—"a Mecca for Microbiologists"—now could be achieved. In June of that year, the Institute of Microbiology opened its doors on the university's rolling campus site overlooking the Raritan River in nearby Piscataway Township, with Waksman serving as director for its first four years. The four-story red-brick Georgian-style building was made possible by a $3.5 million grant by the renamed Rutgers Research and Educational Foundation to Rutgers University. The new institute had well-equipped laboratories and a fermentation pilot plant. It was regarded as the second institution in the world to be devoted exclusively to all the disciplines of Microbiology, the first being the Pasteur Institute in Paris.

Waksman continued as a prolific contributor to scientific literature. A man of incredible industry, he authored over 400 scientific papers, as well as twenty-eight books. It is not surprising that this highly focused scientist titled his autobiography *My Life with the Microbes*.

AFTERMATH

With the success of streptomycin, the pharmaceutical industry embarked upon a massive program of screening soil samples from every corner of the globe. Other soil microorganisms soon yielded their secrets, and it was the actinomycetes that generated a cornucopia of new life-saving antibiotics.

Selman Waksman died in 1973 at the age of eight-five, widely regarded as "the father of antibiotics." An inscription on his gravestone quotes a passage from Ecclesiastes that was a favorite of his: "The lord hath created medicines out of the earth, and he that is wise will not abhor them."

Schatz was never able to gain a position in a first-class microbiology laboratory. In the early 1960s, unable to find work in the United States, he took his family to South America, where he worked as a professor at the University of Chile. His interests became nutrition, pollution, dental caries, and water fluoridation. He returned to the United States as a professor of science education, first at Washington University and then at Temple University in Philadelphia.

As time went on, Waksman continued to be eulogized, and Schatz felt he was being written out of history. He was doggedly determined to voice his side of the affair to as wide an audience as possible, to establish his credit for the discovery of the wonder drug streptomycin. But traditional sources like major academic journals were closed to him. "He assiduously submitted articles on his role in streptomycin to a number of journals," recalls his wife Vivian, "but they were rejected."[77] As rare opportunities arose over the decades, he seized upon them even though they offered nothing more than a voice in the wilderness. The first was an article published halfway around the world in the *Pakistan Dental Review* in 1965 with the title, "Some Personal Reflections on the Discovery of Streptomycin,"[78] followed decades later by a memoir with the title "The True Story of the Discovery of Streptomycin" in the equally obscure journal *Actinomycetes*.[79] Unsurprisingly, both these efforts were of little consequence, but they do spotlight Schatz's unceasing distress. Then, in 1991, Schatz's cause was taken up by Milton Wainwright of the Department of Molecular Biology and Biotechnology of the University of Sheffield in England.[80] Wainwright reviewed the details of the conflict and interviewed Schatz in his home in Philadelphia over four days. "Wainwright was also of working-class origin," Vivian Schatz says, "and was sympathetic to the story told by Albert, who often was in tears as he recalled the painful memories."[81] Clearly touched by his plight and minimizing or overlooking several pieces of archival evidence that support Waksman's perspective, Wainwright served as the key to Schatz's rehabilitation. In short time, the Theobald Smith Society (the New Jersey branch of the American Society of Microbiology) awarded Schatz its most prestigious prize, ironically named the Waksman Award and Medal. And in 1994, at a ceremony celebrating the fiftieth anniversary of the discovery of streptomycin and hosted by Rutgers' president, Albert Schatz was presented with the Rutgers Medal, the university's highest award. This recognition gave Schatz great personal satisfaction and a feeling of vindication.

Instrumental in bringing this about was Douglas Eveleigh, a professor of microbiology, a fair-minded, affable, and energetic Englishman on the Rutgers faculty since 1970. As the "keeper of the flame" for preserving the institutional memory of Selman Waksman, Professor Eveleigh established a permanent exhibit in the basement laboratory where streptomycin was discovered, in what is known now as Martin Hall, a red-brick Georgian-style

Figure 6.7   Albert Schatz, age 72, in 1992. [VS]

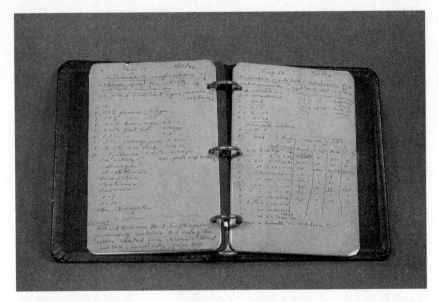

Figure 6.8   Selman Waksman's notebook describing experiments establishing the effectiveness of streptomycin. Added on an unknown date as a postscript in this entry dated 9/15/43 is the comment: "[Regarding] the 2 streptomycin producing cultures, D-1 being the culture isolated from chicken's throat and 18-16, 2 days later, from soil." Waksman contributed this notebook to the Smithsonian Institution in 1953. [SI]

building on the university's College of Agriculture campus in a gracious setting of a lawn and grove of trees. Designated a national historic landmark by the American Chemical Society, a viewer can marvel at the ingenuity of self-made instruments, the striking display of actinomycetes, and notebooks of precise observations.

Albert Schatz died in June 2005 at the age of eighty-four. His case had become widely championed in books[82] and articles.[83] To the end of his life, Schatz maintained a sense of grievance and never ceased to argue his case for recognition. In a small book published the year after his death, he continued to defend the legitimacy of his claim.[84]

CHAPTER 7

SHOCK WAVES IN ACADEMIA

The noblest ambition is that of leaving behind one something of permanent value.

—*G. H. Hardy,* A Mathematician's Apology

How are we to judge this conflict from the vantage point of over half a century later? The allocation of credit for the discovery of streptomycin is one of the most contentious issues in medical history. It is emblematic of the struggle for recognition and reward in scientific discovery. Beyond these issues are the elements of the human drama that sparked such an explosive rupture. Tracing the perceptions and judgments, often false and inaccurate, of both Schatz and Waksman, unveils a Rashomon-like quality in which each pursued a path of self-aggrandizement. The detailed archival material of each reveals the steps leading to an inevitable clash, made all the more acrimonious because it was initially founded on a relationship of mutual respect and admiration. Men of scientific integrity were not above employing ruses to further their prerogatives. Distortions and rationalizations, even unintended consequences, led to embittered self-righteousness.

To Albert Schatz, his senior authorship of two papers announcing to the medical world streptomycin and its effectiveness against the tubercle bacillus, his listing as co-inventor with Waksman on the patent application, and his Ph.D. thesis on streptomycin clearly earned him the distinction of being acknowledged as the co-discoverer, no less the discoverer, of this new wonder drug. In his mind, his priority claims were clear:

> Waksman became interested in and involved himself in my research only after I had isolated the two strains of *S. griseus*, demonstrated that they both produced the same antibiotic, established that streptomycin...was a new antibiotic, identified *S. griseus*, and found that streptomycin inhibited the growth of the tubercle bacilli *in vitro*.[1]

Schatz was motivated primarily by an intense drive for recognition. This was the principle by which he led his life, to be "true to my own convictions." He was oblivious to the practical consequences of bringing a lawsuit against Waksman but came to despair the hopelessness of evading censure.

Schatz's seniority on the seminal papers with Waksman as a co-author is a remarkable exception to the published output from the laboratory. Schatz believed it to be an open admission of his major contribution. Waksman explained that it was an act of generosity on his part, typical of his practice to give seniority on papers to his graduates in order to advance their careers. But a review of Waksman's list of prior publications shows that his almost unvarying practice was to put his name first on research papers originating from his laboratory.[2] Interestingly, Doris Jones's work with chick embryos also earned her senior authorship on the publication dealing with streptomycin's *in vivo* effectiveness against tuberculosis,[3] strengthening the suspicion that Waksman decided to relegate his position to a secondary one on the laboratory's initial publications on the tubercle bacillus. Only thereafter did he, in his capacity as the famed director of the antibiotic program at Rutgers, assert his prominence with streptomycin.

Schatz's sense of outrage must have been monumental when he perceived that he had been misled into signing away his legal rights to streptomycin. Only fifteen months after filing the patent application, Waksman asked him to join him in assigning the patent to the Rutgers Research and Endowment Fund for one dollar. This followed immediately upon Merck's relinquishing its exclusive rights for commercialization. It was not until the lawsuit in 1950 that Schatz learned that Waksman and Rutgers had been operating behind the scenes with the treasure lode, without disclosure to him.

In Waksman's view, the discovery of streptomycin was a logical culmination to his career as a microbiologist strongly committed to pursuing practical applications for his science. As a product of nature, *Streptomyces griseus* could not be granted a patent, whereas the modifications carried out to purify and stabilize it amounted to the production of "a new composition of matter" as required by the Patent Act. Thus, the wording of the Nobel Prize to Waksman for his "ingenious, systematic and successful studies of soil microbes that led to the discovery of streptomycin" described his achievement precisely.[4] Waksman had established a painstaking screening process along the lines of the systematic testing techniques that had made the German pharmaceutical industry so successful. Schatz always strenuously maintained that it was his determination and energy that led to the discovery of streptomycin and that Waksman diabolically marginalized his role as it became apparent that the drug would be a major breakthrough in the treatment of tuberculosis. But in truth the work could have been carried out by any of several students. Talent and intelligence, along with tireless hard work, helped Schatz toward the goal, but luck placed him with the right project at the right time. As the research had become a matter of large-scale, methodical screening, little scope was left for initiative or individual creativity.

Schatz was in the laboratory for only about three months in 1943 before he isolated the strain of *Streptomyces griseus*. That he achieved this "so quickly is one of those rare quirks of fate which most scientists merely dream about but never experience."[5] Although he maintained that Waksman was away for much of this time, Elizabeth Bugie observed that Schatz's very frequent trips up the three flights of steps from the basement laboratory to Waksman's office on the top floor became something of a joke among members of the department.[6] When H. Christine Reilly, another graduate student, was asked, during the pre-trial hearings, what Waksman's role was in the writing of the first streptomycin paper with Schatz as senior author, she replied forthrightly, "He furnished the brains."[7] Boyd Woodruff commented that Schatz made a chance discovery: "All he had done was a routine screen,"[8] continuing the general research approach he had pioneered in the discoveries of actinomycin and streptothricin. Woodruff had been credited as co-author with Waksman in publications and a partner in patent applications on these drugs, and Waksman had every expectation that he would be able to oversee their development and commercialization until their proven toxicity made this unfeasible.

Waksman did not see his own actions as duplicitous or a betrayal. He genuinely believed that he deserved the credit and reward based on years of effort. Even with his view that Schatz was "one of many cogs in a great wheel," that his students and assistants "were my tools, my hands," there was no malice on his part. He was influenced by the European hierarchical tradition in which credit generally goes to the head of the department. He was, simply, a man of his time. It follows from this view, shared by many senior researchers, that Schatz would be seen by the scientific community at the time not only as a renegade but as an undeserving, inappropriately aggressive, and destructive ingrate.

Receiving Schatz's frequent spontaneous declarations of his total disinterest in money ("I have not the slightest desire for fame, glory, popular acclamation, or a lot of money" and "I simply would not know what to do with more money if I had it"), Waksman could not but be lulled into a perception of Schatz as a researcher immersed in his work without regard for the development and commercialization of any discovery. In fairness, both of these statements were made when Schatz was in despair over his jobs and his career was not going well. Little did Waksman understand Schatz's compelling need for recognition. This was an intense young man who had signed the patent papers as a "co-inventor" in the faith that it would prevent monopolization and that no one individual would profit personally.

Waksman's vision, on the other hand, was to make a lasting footprint with the Institute for Microbiology and the Foundation for Microbiology.

Waksman did not leave to chance the selection of a biographer; he published his memoirs in 1954 when he was sixty-six. It gave him an opportunity not only to recall his multiple contributions but to lay out his self-vindication as well. He never referred to Schatz by name but only referred to him as "a student."

In a speech that same year about the satisfaction of a career in research, Waksman chose to reflect on the tension between the dilemma of attribution and personal ego. "The question frequently arises of how to distribute the credit for a new discovery, a new fact, a new observation," he posited. "Some investigators will credit their assistants merely by footnotes on scientific papers."

"Others," he continued, clearly thinking about his own stance over discoveries in his laboratory, "will exert every effort to encourage their students, inspire them, and stimulate them further to select research careers, and in doing so, may not only add their names to the scientific papers but even place them first. In most cases, the results of the second practice are entirely satisfactory. Often, however, the consequences are most unexpected. How should one strike a happy medium? Scientists are searchers of truth. But they are also human beings."[9] Waksman could take little comfort from the axiom that no good deed goes unpunished.

Later, in an unguarded moment, Waksman offered a candid explanation of why he had not simply answered Schatz's questions at the beginning and resolved the problem immediately. The questions Schatz was asking, Waksman felt, were none of his business.[10]

Assuredly, the matter had been clumsily handled by Rutgers. Nevertheless, it seems likely that a few words of explanation and recognition by Selman Waksman to Albert Schatz, rather than marginalization, would have done much to reconcile the widening break and render a lawsuit needless. Waksman's son, Byron, a crusty personality, had always been a staunch defender of his father's reputation and was embittered toward Schatz. Yet, in an interview in 2008, at the age of eighty-eight, he acknowledged to me that he is "softening my attitude and beginning to understand Schatz's reactions."[11]

Human emotions, human ego, not science, drove them apart. Both were right. Both were wrong.

And therein lies the tragedy.

A HOLLOW VICTORY

Waksman's 1964 book was optimistically and prematurely titled *The Conquest of Tuberculosis*[12]; since then the white plague has returned with a vengeance.

In 1952, the same year that Waksman received the Nobel Prize, his discovery was eclipsed by the introduction of the chemical isoniazid (INH), a significantly more effective treatment for tuberculosis that could be taken orally. With the further discovery that drugs given in combination increased efficacy, the annual death rate from tuberculosis per 100,000 people in the United States fell from 50 to 7.1 by 1958.

However, by the end of the 1970s, strains resistant to all major anti-tuberculosis drugs emerged, and in 1973, the World Health Organization (WHO) declared tuberculosis, which claimed about 2 million lives each year, a global emergency. It is especially rife in developing countries. It ravages particularly the poor, overcrowded, and malnourished, and those whose immune systems have been weakened by HIV/AIDS.[13] The WHO estimates that by 2020, one billion people will be newly infected and that 35 million will die from tuberculosis if control is not further strengthened.[14]

CHAPTER 8

"THIS SHAMEFUL WRONG MUST BE RIGHTED!"

The institution of science puts an abiding emphasis on significant originality as an ultimate value, and demonstrated originality generally means coming upon the idea or finding first.

—Robert Merton, "Priorities in Scientific Discovery"

THIS SHAMEFUL WRONG MUST BE RIGHTED!

On October 10, 2003, readers of the *New York Times* were startled to come across a full-page advertisement with this declaration trumpeted in big bold letters. As if any question of injustice remained, prominently pictured at the top of the advertisement was an upside-down prize medal with Alfred Nobel's image. At issue was the announcement four days before by the Nobel Prize Academy that the award in physiology or medicine was going to Paul Lauterbur of the University of Illinois at Urbana-Champaign and Sir Peter Mansfield of the University of Nottingham in Britain. The $1.3 million-dollar award was given for their work that "led to the applications of magnetic resonance in medical imaging." MRI (magnetic resonance imaging), by this time, had become a household phrase.

This was the first of eight such advertisements, which continued to be published through November and the first week of December in the *New York Times*, the *Washington Post*, *Los Angeles Times*, *Times of London*, and in *Dagens Nyheter*, Sweden's second-largest morning newspaper.

The advertisements were attributed to "the friends of Raymond Damadian," but it was clear that Damadian himself—a physician-researcher turned entrepreneur who had made original and fundamental observations that generated advances by others—had rapidly organized a public campaign to have the decision overturned and advance his claim to sharing the prize.

In the ninety-eight-year history of the Nobel Prizes, such an attack was unprecedented. Disagreements and resentments may have been stirred in the past over certain awards, but there had never been such a public denunciation coupled with a passionate appeal for emendation, a course the Royal Academy has never taken. The awards committees make their decisions in deep secrecy, and their deliberations are sealed for fifty years. The ads asked the public to petition the Nobel committee to include Damadian in the prize. One appealed directly to the doctors on the Nobel committee.

"My main motivation," Damadian explained to me, "was to express my distress at the gross miscarriage of justice. Except for what I did, there would be no MRI today. Probably I had some remote hope, a faint expectation that it would be rectified."[1]

In Stockholm, Hans Jornvall, secretary of the Nobel Assembly at Karolinska Institutet, which picks the winner in medicine, noted that a Nobel Prize could not be appealed. "This is the first time I have heard of somebody taking out an advertisement," Jornvall said. "Science is my life. I would like it always to be happy but this situation is clearly not happy."

As we will explore, Damadian's claims are based on several of his seminal contributions: the original conception of using magnetic resonance (MR) for whole-body screenings of living humans; the fundamental discovery of MR differences among normal tissues and between normal and cancerous tissues, which provides the biological basis for MRI; and the construction of the first full-body human MRI machine, albeit using a crude scanning method.

As in the decades-long conflict between Waksman and Schatz, the thirty-year dispute between Damadian and Lauterbur illuminates two men who in their similarities, differences, and self-contradictions embody the conflicts inherent in scientific discovery. How is credit established? By what measure is priority determined? Is scientific research as driven by elements of human character, particularly ego, as many other endeavors? Does scientific rivalry drive the development of a life-saving technology?

An Innovative Tool

MRI use is now so widespread that everybody seems to know somebody who benefited from the study. Today MRI is a billion-dollar industry, with thousands of MR scanners and more than 60 million scans performed worldwide each year for medical diagnosis. MRI relies on the magnetic properties of the hydrogen atom in water, which accounts for about two-thirds of the human body weight. There are differences in water content among tissues and organs; in many diseases, the pathological process results in changes of the water content. This is reflected in the MRI picture.

When the hydrogen atoms are placed in a powerful magnetic field and bombarded with radio waves, they emit radio signals that provide information about their local environment. Essentially, MRI turns hydrogen atoms in the body into tiny radio transmitters. By tracking where those atoms are, MRI builds up a picture of internal organs and structures. MRI provides a contrast between tissues and disease over forty times greater than the X-ray. MRI is particularly superb in visualizing soft tissues in the human body in meticulous detail, for example, the brain and spinal cord, joints and their cartilages and ligaments, the abdominal organs, and bone marrow that could not be seen by X-ray. Further, in contrast to CT scans, the technique is not impeded by bone so that it displays minute differences in underlying tissue density, and it does not use radiation.

MRI has been so successful that the original technique has spawned numerous offshoots. Functional MRI identifies which different parts of the brain are active during different activities by identifying where the most blood is flowing. It permits study of the emotional and behavioral centers

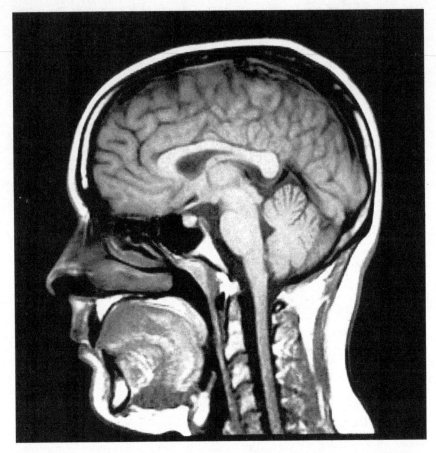

Figure 8.1   MRI of the brain and spinal cord showing precise anatomic detail.

in the brain and the effects of drugs upon them. Diffusion MRI can distinguish between tumors and the surrounding region of edema and can detect strokes by measuring the movement of water across microscopic distances in the brain. MRI angiography diagnoses heart disorders by taking pictures of blood vessels.

The strength of magnets is measured in gauss and Tesla units. Today, clinical imaging is done at a magnetic strength of 1.5 or 3.0 Tesla. The magnetic field of the Earth is half a gauss, and there are 10,000 gauss to a Tesla. The most powerful magnets produce images with the highest spatial resolution. A 9.4 Tesla MRI unit at the University of Illinois at Chicago—the most powerful in the world for human use—has the power to go beyond imaging of hydrogen to imaging a palette of elements, such as carbon, sodium, phosphorous, oxygen, and nitrogen. This shifts the function of MR from anatomic to metabolic. Metabolic imaging enables scientists to look at the expression of genes and the functions of the cellular machinery, providing a new biochemical dimension of medically relevant information.

AN UNEXPECTED SPIN-OFF

MRI is an outstanding example of a completely unexpected spin-off from a basic science research tool, nuclear magnetic resonance (NMR). NMR spectroscopy was developed over a span of fifty years by physicists, chemists, and engineers, primarily to solve the structures of organic molecules. It is a technique that exploits the magnetic properties of certain nuclei. The most important applications for the organic chemist are proton NMR and carbon-13 NMR spectroscopy.

Protons and neutrons form an atom's core, or nucleus, while electrons—carrying electrical charges—orbit nearby. These nuclear particles spin and create magnetic forces within the atom. Physicists discovered that the nuclei of certain atoms, when placed in a static magnetic field, will line up like compass needles in the direction of the field. They spin at a frequency determined by the strength of the field. The resonance part of NMR occurs when radio waves at the same frequency—one matching that of the nuclei—are applied, and the nuclei become "excited" and absorb additional energy from those waves and are nudged out of alignment. When researchers turn off the second field, the nuclei wobble around the static field like spinning tops, a type of motion called precession, returning to their original energy level. In the process, they expel the tiny excess energy as radio wave signals that scientists can detect and analyze to understand atomic structures. The relaxation times—the period that it takes the nuclei to flip back to their original orientation—are manifested in two forms, called T1 and T2. NMR provides information on the number and type of chemical entities in a molecule, on mixtures, and on dynamic effects such as change in temperature and reaction mechanisms, and it is an invaluable tool in understanding protein and nucleic acid structure and function.

NMR has an illustrious history, garnering a succession of Nobel Prizes by physicists and chemists. I. I. Rabi at Columbia University received the award in 1944 for his discovery that the nuclei of atoms are magnetic. This set the stage for the award in 1952 to physicists Felix Bloch at Stanford and Edward Purcell at Harvard for their independent demonstrations that it is possible to obtain chemical information about the internal structure of molecules, and NMR machines were built to exploit this phenomenon. In 1991, Richard Ernst, a professor at the University of Zurich, received the Nobel Prize for chemistry for advancing the field of NMR further and making it the sophisticated, analytic method of today, followed by the award in 2002 to Kurt Wüthrich, also from Switzerland, for his contributions to two-dimensional NMR spectroscopy.

But nobody thought of using NMR for imaging biological tissue because researchers were long focused on chemistry and simply didn't know how it could be done. It was Raymond Damadian who first envisioned NMR's potential as an enormously valuable medical tool and attracted the attention of the medical community.

Damadian's claim on priority for biological applications is based on his groundbreaking observation of significant differences in relaxation times, T1 and T2, between normal and malignant tissue in rats. Organs and tissues have distinct relaxation times, and Damadian found that cancerous tissue takes longer to relax than healthy tissue. This sparked the development of the concept of an MRI scan that translates the relaxation rates into levels of brightness and then yields a computer-generated image of the organs and any tumors.

It is important to understand the chronology of events that contributed to the move from the chemistry or physics lab to the bedside or clinic in order to understand the ongoing controversy.

WRITTEN OUT OF HISTORY

The long controversy regarding the development of MRI, some felt, kept the Nobel committee from acting earlier. Lauterbur, 74, was in declining health, however, and the Nobel Prize cannot be awarded posthumously. If the contribution of MRI was to be acknowledged, as Richard Ernst urged it must be, the time had come.

For the past ten years, Damadian had been actively lobbying the Nobel committee, providing loose-leaf documents with letters of endorsement from eminent figures, including several Nobel laureates, and magazine articles and books detailing his contributions to the history of MRI. "Their decision," he ruefully told me, "doesn't occur accidentally at night. I heard they were thinking of giving the prize to the MRI and I wanted to be sure they understood that it was my idea."[2]

Most letters to the Nobel committee testified solely to Damadian's discovery, but, interestingly, a few recommended that the prize should be shared by Damadian and Lauterbur, and one went further to dispel any bias generated by Damadian's conduct: "[Their] contributions are so interrelated that it would be an injustice to single one over the other. I also trust that Dr. Damadian's…idiosyncrasies will not prove to be an obstacle, as many brilliant and productive scientists do sometimes exhibit unorthodox behavior."

As a major step in averting "being written out of history," Damadian supported, if not commissioned, an impressive 838-page volume on the history of the discovery and the development of NMR and its modern applications in medicine.[3] The act can be viewed as one of either forceful initiative and resourcefulness or shameless self-promotion. The book profiles nine pioneering scientists from the fields of physics, chemistry, and medicine, including such luminaries as I. I. Rabi, Edward Purcell, Felix Bloch, and Richard Ernst. Unsurprisingly, Raymond Damadian shares a place in this pantheon, as does Paul Lauterbur.

In addition, during this time, two distinguished Swedish physicians involved with diagnostic imaging had visited Damadian separately: Hans Ringertz, a self-assured pediatric radiologist and scion of a distinguished family, and Bjorn Nordenstrom, a retired thoracic radiologist widely respected for his judgment. Were they on a mission from the Nobel committee? According to Damadian, they did not present themselves as the committee's emissaries of the Nobel committee.

But his hopes were dashed in the early morning of the sixth of October. Aware of international time differences, the Nobel committee informs the winners by telephone before announcing them to the media. At 3:30 A.M., CDT Paul Lauterbur's phone in Illinois rang, and he was prodded awake by his wife from a deep slumber with the words every researcher dreams of: "It's Stockholm calling." Damadian got no such call.

Two hours later, the Karolinska Institute announced to the world that Lauterbur and Mansfield had won the 2003 Nobel Prize in physiology or medicine, deemed the most coveted and celebrated honor of human achievement, "for their [seminal] discoveries concerning magnetic resonance imaging." The wording, with its emphasis on imaging, was carefully chosen.

Alfred Nobel's will allows up to three individuals to be honored and specifies only that the selection be for the discovery that has "conferred the greatest benefit on mankind."

In the pre-dawn darkness in Long Island, Raymond Damadian logged on to the Internet, as he had done in recent years, and was, in his words, "in anguish. I thought: 'No, I cannot live with this.'" It was like a knife in his heart to see not only that he had not been named but that the third position was simply left open. To Damadian as well as his supporters, the omission was clearly purposeful.

Damadian is a notorious figure among scientists because he has courted public attention and controversy for over thirty years. He has continuously maintained that it was he who discovered the crucial first step that led to the development of MRI and continued to make other landmark achievements. He sees himself in the grand tradition, as a maverick researcher, an outsider, who brought his intuitive mind to a new discipline to ask and answer questions about the "big picture" with enormous potential.

Damadian had long feared that his contributions were "being written out of history," and his bitterness and antagonism simmered for three decades with occasional volcanic eruptions. He harbored intense personal hostility against Lauterbur in particular.

In an interview with the Sydney *Morning Herald* a week after the announcement, Damadian lamented, "I don't understand why [Lauterbur] would want the prize when he knows that I am not sharing it with him, and he knows the idea was mine. And there is a judge that he is going to have to answer to, and so I grieve for him."[4]

He considers the contributions of Lauterbur and Mansfield to be only technological refinements and emphasizes that "the idea of an MRI didn't occur to [them] until after a medical doctor," speaking of himself, "made his landmark discoveries." Throughout, he disparages "the insider club of NMR chemists and physicists" for not being willing to accept that "a medical doctor made the [breakthrough] discovery in their field." In his mind, he had opened the door, others rushed to cross the threshold, and he was left ultimately holding the doorknob. Lauterbur, who kept himself generally aloof from the priority dispute, haughtily remarked to a reporter that Damadian "demonstrates the kind of behavior that is not regarded as responsible in scientific circles."[5]

Damadian's newspaper advertisements overflow with unbridled outrage and at times wrathful righteousness. The sense of deep hurt is expressed as flagrant hubris: He refers to himself as "a man who is, because of his brilliant mind and indomitable spirit, one of the greatest living benefactors of patients worldwide and the last person who deserves to be needlessly injured by a prize committee or by anyone else." He excoriates the Nobel committee for "inexcusable disregard of the truth," "political chicanery," and being "ethically and intellectually corrupt." He emphasizes his claim that "the two winners acknowledge that their work grew out of Dr. Damadian's prior discoveries in magnetic resonance" and in a last desperate measure demands that they put aside any enmity and embrace him as a co-winner: "The time has come, not only for great science, but for greatness of character. It's time for one and all to relegate any disputes or ill will to the past, so all three may share the award." In truth, many who had observed Damadian's polemics over decades felt that the "disputes or ill will" were generally on his side.

Dr. Hans Ringertz of the Karolinska Institute officiated at the awards ceremony in Stockholm on December 10. He was chairman that year of the Nobel Assembly and had been widely quoted for having resolutely declared at the time of Damadian's published pleas, "We made the right decision."

Several years earlier, while attending the annual European Congress of Radiology in Vienna, I met Hans Ringertz by chance. The meeting draws radiologists chiefly from European countries but also from the United States and Asia. We were both staying at the same hotel and late at night independently wandered into the hotel's Japanese restaurant seeking a *kaiseki* meal. We were the only customers so introductions led easily to dining together. Much of our conversation centered on the striking advances in CT and MRI featured at the meeting by both participants and vendors.

I casually mentioned that it was a professor of chemistry at my university, Paul Lauterbur, who had devised the technique for generating an image by a process known as NMR, which led in time to the explosive development of MRI. Ringertz was clearly not unfamiliar with this background. He off-handedly mentioned that his father had been dean of the Karolinska Institute, a medical university that, left unstated, was widely known to award the Nobel Prizes in physiology or medicine; his father, Ringertz added, had served on the Nobel Prize committee from 1951 to 1969.

A COMPLEX CHARACTER

I recently interviewed Damadian in Melville, Long Island, in his office at FONAR, the corporation he founded to make, service, and sell MRI machines. With a white mane, full mustache, and prominent nose, he bears more than a passing resemblance to the writer William Saroyan. He was engaging, articulate, and garrulous and projected a physical presence that belied his average height. With a razor-sharp memory, he was able to recall minute details of events over thirty years ago. Stimulated by a query or statement, he typically amplified his response by beginning with, "There's more to it than that..." or "I think it's important to understand that..." These were said not as a polemic but as an earnest attempt to reframe a perspective, to present events through his perceptions.

His gaze occasionally drifted toward a small woven rug with an abstract design in blue, red, and white mounted on the wall near his desk. Anyone familiar with Damadian's publications would recognize it as the first image of a live human body using NMR, accomplished in 1977. But there was nothing wistful in his tone. While sipping green tea, his reflections on the Nobel Prize were somber and philosophical: "I really don't suffer about it at all," he said, adding after a long pause, "I just don't value it. It's lost its credibility." I reminded him that the Nobel Prize is still, however, regarded as the crown of scientific accomplishment. "The Nobel Prize now has the stain of Cain on it," he replied. "I take solace from the Bible (Luke 16:15): 'That which is highly esteemed among men is an abomination in the sight of the Lord.'" His early vision of a whole-body machine was thought by others "bordering on the fringes of lunacy: I had to live with that for a long time." His bitterness toward Lauterbur persisted unremittingly. "Even though he acknowledged my priority in his earliest notes, he buried that fact for twelve years."

Damadian's father was born in Kayseri, a small town in eastern Turkey under the Ottoman Empire, and at the age of twelve narrowly escaped the 1915 Armenian genocide that took over one million lives. When I pointed out from my research that Damadian shares his middle name, Vartran, with his father's first name, and further, that it means "iron shield" (historically, for self-defense against the waves of empires over the centuries), he said, laughingly, "Sure, against Lauterbur!" This was five years after the Nobel Prize and one year after Lauterbur's death.

Damadian long manifested the qualities of a complex tortured character, a tragic hero with Olympian strengths and flaws. Over the years, many in the academic community viewed him as bombastic, brash, bluntly dismissive of criticism, fearlessly provocative, and egotistical. To them, he seemed at times almost paranoid, subscribing to a conspiracy theory that he was being denounced, excluded. This perception would be punctuated by episodes of self-promotion and public rants against the "NMR community" and particularly Lauterbur. Among a small coterie of Damadian's supporters, John Throck Watson, Ph.D., professor emeritus of biochemistry and chemistry at Michigan State University and a friend for the past forty-five years, sees him as a visionary genius, although excessively self-defensive.

Such charges, Damadian stated, may be a "justifiable conclusion if you're sitting outside. Had I not fought back, I believe there would be no MRI today. I was attacked by all the vested interests. I had two choices: pack it in or fight back. I fought back and argued the point." Yet, others observed that Damadian "needed an enemy," that he functioned best when he was in competition or conflict with a perceived adversary. Aggrieved, he could channel his resourcefulness, initiative, and perseverance with focused intensity. Lauterbur, in contrast, was generally of a dispassionate temperament, and many of his colleagues came to believe that in his heart he considered Damadian a charlatan.

Damadian referred to a recent authoritative textbook, *MRI From Picture to Proton,* published by the Cambridge University Press in 2003, in the months just before the Nobel Prize award, which stated, "The initial concept for the application of NMR, as it was then called, originated with the discovery by Raymond Damadian in 1971 . . . This exciting discovery opened the door for a complete new way of imaging the human body." Damadian proudly stated: "I couldn't have said it better myself."

Unknown to Damadian, in the second edition in 2006 (after the Nobel Prize awards), the authors bowed to official judgment and deleted these few lines of tribute. I chose not to inform him that a landmark discovery applauded in early 2003 was no longer cited a few years after the Nobel Prize awards. This observation would certainly have reinforced Damadian's bitter contention that the significance of his contribution was being "written out of history."

In the immediate aftermath of Damadian's published diatribes against the Nobel committee's decision, several voices of disdain were raised. Francis Bonner, a professor emeritus of chemistry at the State University of New York at Stony Brook, who as the department's founding chairman had originally recruited Lauterbur in 1963, remarked to a *New York Times* reporter, "I think it's unfortunate that Dr. Damadian has elected to make such a public fuss. It is not enhancing his reputation in any way."[6] A contrary view was openly expressed by Robert Furchgott, a 1998 Nobel laureate for his work identifying nitric oxide's role in the cardiovascular system. "I don't know why [Dr. Damadian] has been ignored," he said. "I think it's wrong for him to be excluded. Everybody knows that he made a major contribution."[7]

From a Paper Glider to a 747

The odyssey that would carry Raymond Damadian from an incandescent insight through an obsessive pursuit began in 1969 with an interest in an unorthodox proposition—now generally discarded—that water in malignant cells differed in organization from water in healthy cells. Damadian was then a physician scientist in the Division of Biophysics of the Department of Medicine at the Downstate Medical Center in Brooklyn, part of the State University of New York system.

The best place to start to investigate this was at NMR Specialties in New Kensington, Pennsylvania, which lies on the Allegheny River near Pittsburgh. This small company manufactured and sold NMR equipment to academic institutions and corporations. The technology had increasingly gained respect not only among physicists but also chemists and medical researchers, who were interested in the body's chemistry. The company was generous in making its instruments available at the site for brief research projects, since these scientists, along with their departments, were potential buyers. NMR Specialties, distant from hi-tech centers and marketing a niche product, served as a mise en scène for dramatic experiences leading to a revolutionary advance in medicine.

Damadian, with no previous experience with the technique, was astonished and exhilarated to witness the potassium signal on the oscilloscope screen. Part of the NMR spectroscopy unit, an oscilloscope is an EKG-like monitor for measuring electrical impulses in different configurations. He was seized with an insight that would dominate his thinking and actions for the next eight years. He believed he would be able to distinguish cancerous from healthy tissue on the basis of the cell's water structure. Water's composition as H_2O indicates that a molecule contains two hydrogen protons. With the training of a physician as well as a biophysicist, Damadian excitedly jumped to postulating NMR of the complete human body.

Chemists had been employing NMR for over twenty years, but their focus—virtually without exception—was too narrow to envision medical applications. As an outsider coming to the field, a medical doctor with a scientific background, Damadian's huge leap in imagination was a bold change in scale. In Damadian's words, it was like "going from a paper glider that you tossed across the classroom to a 747."[8] At that time, the machine used for NMR spectrometry housed a hollow magnet that could hold a sample no wider than a pencil. To scan even a microscopic sample took a machine that weighed a ton. Scanning a person would require an entirely new apparatus.

Damadian's thoughts raced: "If you could do a human being, scan across the anatomy, to yield detailed chemistry of every organ, you could spark an unprecedented revolution in medicine." Damadian's vision was focused on mapping chemical data and not on imaging. Damadian was undaunted. In September 1969, he applied to the New York City Department of Health's Research Council for a grant and was funded $89,000 to acquire a super-conducting NMR unit. This would amplify sensitivity twenty-five times. His intention was to measure NMR relaxation times in tumors as the basis for an NMR body scanner "for early signs of malignancy." As reflected in the concluding sentence of his letter of application, Damadian maintained throughout his career an unswerving goal of "the total eradication of this disease [cancer]."

But he was way ahead of himself. He had not yet provided any proof that NMR could distinguish cancerous from non-cancerous cells. As often happens in scientific research, he had started on the wrong end of the right path.

Within a few months, Damadian returned to NMR Specialties with a small colony of new rats, while broadening them to include studies of hepatoma (liver cancer). A novice in NMR spectroscopy, he mastered the technique "over the better part of a week." Sacrificing the rats, he tested samples of tumor tissue and was excited to find that hepatoma tissue differed sharply from normal liver tissue. Damadian also found that healthy tissues from various organs—brain, liver, kidney, intestine—showed markedly different relaxation times from one another. This was the first time that noticeable differences in the relaxation times between cancerous and healthy tissues were documented. These differences are the reason that tumors are so brightly illuminated on current-day MR images. On March 19, 1971, the thirty-five-year-old Damadian published the article "Tumor Detection by Nuclear Magnetic Resonance" in the prestigious journal *Science*.[9] Although he did not discuss the formation of images, the article became a cornerstone of what would later be known as MRI. A chasm had been bridged. Enthused Damadian: "I could now distinguish cancerous tissue from normal tissue by the NMR signal." He was encouraged that these distinctive characteristics could provide the potential "of an external probe for the detection of internal cancer." The paper would be designated a citation classic by the Science Citation Index based upon its frequency of citation.

Damadian's article caught the eye of Donald Hollis, a research biochemist at Johns Hopkins Medical School in Baltimore interested in NMR explorations of biological systems. Hollis was concerned that the results might not apply to cancer more generally because Damadian had studied unusually aggressive, fast-growing tumors. Just before the Labor Day weekend in 1971, he dispatched Leon Saryan, a graduate student in his program, to New Kensington to verify Damadian's results in more slowly growing cancers.

ENTER LAUTERBUR

It was at this point that a remarkable chain of chance events brought Paul Lauterbur on the scene. Lauterbur, a forty-two-year-old physical chemist from the State University of New York in Stony Brook on Long Island (only forty miles from the Downstate Medical Center in Brooklyn), had served on the board of directors of NMR Specialties for several years. With the company about to go bankrupt, he was asked to serve as acting president and chairman of the board. Classes were over, he had no grant money to spend right then, and he had no summer salary coming from the university, so he accepted the position.

Tall, with a deep voice, early frontal baldness, and a groomed beard, he viewed the world through large glasses in the fashion of the times, which he would continue to wear throughout his life.

Lauterbur was a productive academic researcher and a leading authority in the field with several innovative articles. Early on, he was introduced to NMR spectroscopy and recognized that it was indispensable for analyzing chemical compounds. He enthusiastically embraced it as his main focus of research. Over the years dating to even before he received his Ph.D. in 1962 at the University of Pittsburgh, he was familiar with the NMR literature, including Damadian's *Science* article. But his interest in NMR had always been that of a chemist, and the idea of applying the technique for medical purposes had never crossed his mind.

He dropped by the company's lab in a large barn-like room to observe Saryan's results. Not used to observing animal experiments, he was shocked to see how the rats were sacrificed and dissected to study the tissues. It was, in his words, "rather distasteful."[10] Saryan would take the animal's tail, twirl the body in the air, and then smash its head against a short piece of railroad track along the edge of a bench before cutting out healthy and diseased tissue samples and measuring the relaxation times.[11]

Lauterbur was struck by the limited nature of the investigation. He shuddered at the idea that "little chunks" of tissue would be cut out of patients in surgery so that NMR measurements could be made while they waited. It made little sense to him to use NMR on excised tumor tissue since conventional histologic examination of biopsy material—stains and microscopic examination of cellular changes—had proved reliable and accurate. What was being presumed was the ultimate utility of an NMR unit in every OR and surgical clinic for on-the-spot studies. Yet, he was impressed by the distinct NMR difference between the various normal tissues and the malignant ones. What would be immensely more useful, Lauterbur mused, would be a method to obtain the same information non-invasively from outside a living body. Was there a way, using NMR, to map out different tissues without cutting into a person?

The thought occupied his mind the rest of the day, and he continued to ponder the question that evening while eating at the "Eat 'N Park," a local fast-food restaurant, with Don Vickers, vice president of applications at NMR Specialties.

Vickers recalled an incident when he generated a signal on an NMR unit to show Damadian on one of his visits. Damadian was unimpressed, saying, "So what! I've seen an NMR signal before." Vickers then revealed that it was produced from inserting his finger in an NMR probe. "My God!" exclaimed Damadian. "You know what this means? We have to build machines large enough to put people in!" According to Vickers, Damadian was all "pumped up over the idea of fashioning it for a valuable clinical diagnostic tool."[12]

As they sat down to dinner, Lauterbur and Vickers began laughing at such a foolish idea because the magnetic field would be so distorted that no reliable data could be obtained. On the second bite of his Big Boy double-decker hamburger, Lauterbur was struck with the solution like a bolt of lightning. In awe, Vickers witnessed the rapid cascade of thoughts that led Lauterbur to his epiphany of how to create images using NMR. Suddenly, Lauterbur said, "Wait a minute! The problem is not that the field would be inhomogeneous, the problem is that we don't know the shape of the inhomogeneity. If we applied field gradients of known shapes, we could get NMR signals emitted from each part of an object so that they would disclose their spatial location. We can make pictures with this thing!"[13]

Normally in order to keep the NMR field homogenous, the samples were spun at high speeds in order to expose them to a uniform magnetic field. Chemists went to great pains to create a uniform magnetic field, under which the molecules gave the clearest signal. If this were not done, the resulting spectrum would exhibit weird shapes and spikes. It suddenly occurred to Lauterbur that these potentially conveyed much hidden and valuable information about the spatial distribution of molecules that was being overlooked and simply not taken advantage of. There was a mother lode of information. All he needed to do was puzzle out how to mine it. But the animal experiments showed that different tissues turn uniform magnetic fields into a number of different local fields, which consequently fire out radio signals at different frequencies.

As the major constituent of water, hydrogen is the most prevalent element in the human body. Lauterbur's epiphany was the basic realization that he could use NMR to produce images by mapping the location of hydrogen nuclei in the body.

He sketched out his thoughts—MRI's first model—on a paper napkin. Up until this time, variations in the strength of the magnetic field were the bane of NMR, an undesirable feature to be obliterated at all cost. With tremendous insight and intuition, Lauterbur turned magnetic field gradients—one of the most problematic issues in NMR—on their head.

Excitedly, he bought a small tan spiral notebook at a drugstore. He spent much of that night refining his thoughts. Before he fell asleep, he had crystallized his concept and neatly written his proposed theory. In three pages, he outlined his approach of using a variety of linear magnetic gradients with different orientations to elicit different radio wave frequencies—that is, relaxation times—from which an image could be computed. Lauterbur clearly envisioned that this "should be capable of providing a detailed three-dimensional map of the distributions of particular classes of nuclei...within a living organism."[14] Lauterbur's brief notes cite Damadian's 1971 *Science* article, a point that would later garner a lot of attention.

As scientists do, to ensure rights to a patent, Lauterbur had Vickers sign and enter the date—September 3, 1971—on his notes as a witness.

Lauterbur returned to Stony Brook for the fall semester, as another board member of NMR Specialties took over the responsibilities of the company, which shortly collapsed, and he set out to test and work out his ideas. The best NMR machine on the campus was located in the chemistry department. Lauterbur experimented on it at night, altering the settings to give the varying magnetic fields he needed, carefully restoring the settings for a uniform field each time before he left.

Designated a national historic landmark by the American Chemical Society in 2011, the original unit is on permanent display in the lobby of the graduate chemistry building on the campus of the State University of New York at Stony Brook. To view the rather unimpressive three-piece machine now—an NMR unit the size of a small refrigerator, a console for controlling the factors, and an oscilloscope screen—is to admire the giant step taken

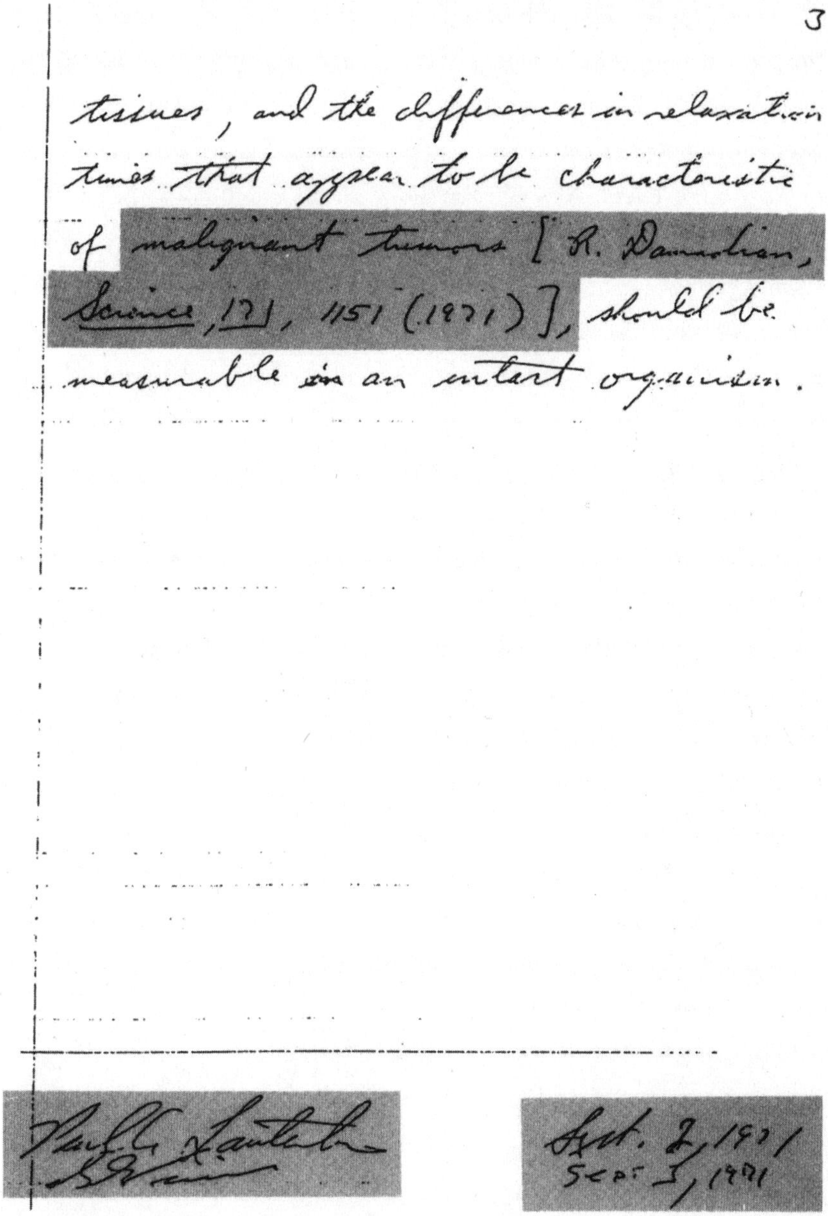

3

tissues, and the differences in relaxation times that appear to be characteristic of malignant tumors [R. Damadian, *Science*, 171, 1151 (1971)], should be measurable in an intact organism.

Paul C. Lauterbur

Sept. 2, 1971
Sept. 3, 1971

Figure 8.2 Paul Lauterbur's notebook entry of September 2, 1971, citing Raymond Damadian's earlier article in *Science*. [DBC]

from the conventional technology of the day. Slowly but surely, Lauterbur evolved a technique of applying magnetic field variations, or gradients, at several defined angles around an object, and recording the different sets of NMR signals at each specific variation.

Lauterbur figured out how to create a series of one-dimensional projections by changing the orientation of the gradient field and then, with the help of a physicist skilled in mathematics from the nearby Brookhaven National Laboratory, how to reconstruct a two-dimensional image from a mass of data in the form of "slices." Several years ago, I asked Lauterbur how he felt about coming up with the key to solving the imaging problem. With his typical crinkly-eyed smile, he simply concluded, "Everything fell into place!"[15] The exhilaration in this statement conveys multiple parameters: an incisive and receptive mind, the joy of insight, the aesthetic elegance of a synthesis, and the introduction of a new view of the universe. He knew that he had reached a turning point: "I had stumbled across an idea that was worth spending most of my time on for the foreseeable future instead of carrying on with molecular structure studies."[16] He places his insight among "the unexpected breakthroughs that grow in the fertile soil of 'curiosity-driven' research," adding that this is "the most cost effective kind in the long run."[17]

Lauterbur was not so starry-eyed that he failed to anticipate NMR's potential for practical applications. Even as he was developing his technique, he also concentrated his efforts on a delicate balance between two major pursuits: publication in a prestigious journal with a wide readership and the security of inviolable patent rights. The first would clearly establish his scientific priority, and the other would legalize his intellectual property rights. He was fully aware that any patent was his exclusively, because the technique was a summer epiphany and not discovered on university time.

To trace the chronology of these two efforts is to raise questions regarding motives and the truthfulness of attributions.

By the end of October, Lauterbur was in contact with Edward D. Welsh, a patent attorney based in Pittsburgh who provided counsel to NMR Specialties. In a letter to Welsh dated November 10, 1971, Lauterbur raised and dismissed the idea of potentially useful and profitable collaboration with Damadian: "Damadian and I neither like nor trust one another."[18] He was referring to his first contact with Damadian a year and a half to two years earlier, which established an immediate distrust that would grow to a bitter relationship over the next thirty years. At that time, if his $89,000 grant application in September 1969 were accepted, Damadian planned to buy a superconducting NMR spectrometer from NMR Specialties. In an apparently bull-headed manner, according to Vickers, Damadian called Lauterbur and said he was listing him as a reference on the grant application, and "since you are on the board of NMR Specialties, I expect a favorable review." Vickers comments: "Everything went downhill from there."[19] Besides, the company was faltering financially and was not in a position to produce the magnet. For his part, Damadian views Lauterbur's rejection as deliberate sabotage born of a competitive impulse and the first of a long history of denials by the "NMR establishment," outright theft, and obstructions to his pursuit of a clinically useful instrument.

CHAPTER 9

THE RACE IS ON

Long flat intervals. Steep, sweaty, even competitive climbs, an occasional cresting of a mountain pass, with the triumphal downhill coast. Always work. Sometimes pain. Rare exhilaration. Delicious fatigue and well-earned rests.
—Harold Varmus, Nobel laureate in medicine, likening scientific research to his avocation as a long-distance bicyclist

Convinced in his bones that he was on to something big, Lauterbur submitted a paper to the journal *Nature* the same month he first contacted a patent attorney. His proof was elegantly simple. The two-dimensional NMR image he used showed cross-sections of two thin glass capillaries of water. For the first time, an image had been produced by NMR, one that would be reproduced in every account of the history of MRI. This seminal NMR image was composed of two irregular circles of cross-hatched lines. Today, various techniques exist to enhance the clarity of an image produced by MR. These are generally referred to as "image processing." The basic parameters essential for quality images include signal-to-noise ratio, contrast-to-noise ratio, and spatial resolution. Lauterbur had resorted to the simple expedient of using a pencil eraser to "smooth out" the fuzzy picture. Observing this crude attempt at "image processing," Waylon House, a plain-spoken postdoctoral fellow in Lauterbur's lab who soon became his research associate, recalls sardonically that at that moment, he lost his naiveté regarding the integrity of "the scientific method."[1]

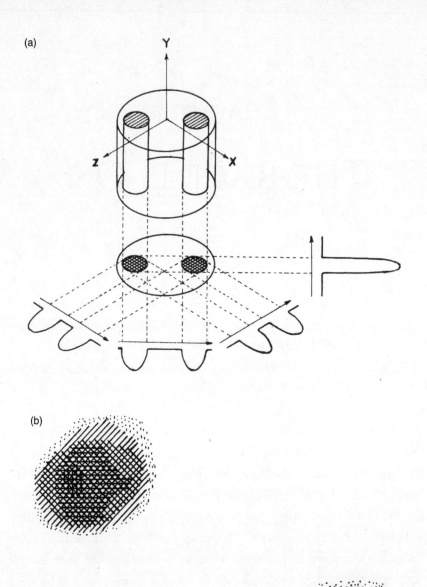

Figure 9.1 First images published by Paul Lauterbur that were produced by NMR:
a. Two glass tubes of water in a field of magnetic gradients.
b. Spatial image of protons. [N]

But the paper was summarily rejected by the editors as "rather trivial" and "not of sufficiently wide significance."

Most journals strive to conceal both the author's and the reviewer's name. This is intended to maintain the objectivity of peer review. Of interest is the fact that although the manuscript's reviewer has remained anonymous, Lauterbur's authorship was clearly evident. The reviewer chose not only to not recuse himself, but his comments were made available to Lauterbur: "The present application is a very trivial one... [Being] aware of Professor Lauterbur's eminent reputation... I am content to accept this, but the journal's editors might like to see more evidence of the usefulness of the technique."[2]

Lauterbur, convinced of his concept, appealed to the editors, noting, "My manuscript announces a new principle of image formation, with innumerable potential applications in science, medicine and technology." He noted that he had used NMR to show the "soft tissues of a clam inside his closed shell" and unboundedly went further to state that "the machine could demonstrate that the features of tissues characteristic of malignant tumors could be detected inside an intact organism."

The atmosphere was ripe for furthering cancer research. President Richard Nixon signed the National Cancer Act into law on December 23, 1971, and announced an all-out federally funded "war on cancer." Without attribution to the work of others, and thus perhaps implying that it was ongoing work in his lab, Lauterbur went on to mention that NMR could also determine "flow rates in hidden blood vessels... and various enzymatic processes."[3] He didn't intend to incorporate these ideas in his revision, but he presented them as persuasive points. This discretion would enable him to secure patent rights.

His revision was published in the March 16, 1973, issue of *Nature*,[4] with the added conclusion that his technique "should find many useful applications in studies of the internal structures, states, and compositions of microscopic objects," surely one of the most understated taglines in scientific literature.

The editors had chosen to delete the words "and macroscopic" that Lauterbur had written after "microscopic" in the tagline, perhaps because they were unconvinced by the fuzziness of the images and by the prospect of applying his method to macroscopic imaging. And yet, he has acknowledged that, "I still did not know exactly *how* it would be useful in medicine to have a method for making three-dimensional pictures of the body, since I knew nothing about medical science or clinical practice."[5] Nothing more about applicability was indicated, even though the document he sent to the patent attorney listed several useful technical refinements of his method of imaging. Lauterbur's article certainly established "a new principle of image formation," and he obviously had determined not to make his technique completely public by publishing before he obtained a patent. He has affirmed that he did not then know of Damadian's "speculative patent application."[6]

Thirty years later, *Nature* counted this work as one of the twenty-one most influential scientific papers of the twentieth century. Lauterbur ruefully remarked that the whole history of modern science might be written on the basis of papers turned down by academic journals.

PROPOSED BREAKTHROUGH SIMPLY STATED AND ITS POTENTIAL UNAPPRECIATED

In the end, science is the search for simplicity. Lauterbur's groundbreaking article was composed of only 840 words. Its brevity was not a flaw. To reduce an issue to its elemental truth is both an art and a creative skill. Picasso at an exhibition of children's art reflected: "When I was their age I could draw like Raphael. But it took me a lifetime to draw like they do." Richard Feynman believed that one should be able to accurately reduce an idea so it can be presented to a college freshman class. Nobody expressed this more clearly and eloquently than Einstein: "Things should be made as simple as possible, but not any simpler." In science, prolixity and the number of co-authors bear no relationship to the substance of a presentation. This is spectacularly illustrated by major contributions to scientific knowledge:

- Wilhelm Röentgen reported the discovery of X-rays in ten pages in the 1895 volume of the annals of his local physics society.
- Albert Einstein developed the special theory of relativity in thirty pages with little mathematics, few footnotes, and no references to other authorities.
- James Watson and Francis Crick described the double helical structure of DNA in 943 words.

- Gerald Edelman explained the composition of immunoglobulins in 384 words. Edelman proudly said, "That paper had a one-sentence entry and one-sentence conclusion, had an equation in between and a table of data, and it changed the nature of the subject."[7]

Lauterbur's advance, so succinctly stated, would move science from the single dimension of NMR spectroscopy to the second dimension of spatial orientation—which proved to be the foundation of MRI. He coined the term zeugmatography—from the Greek *zeugma*, or yoke—for his new technique because it tied together two different types of energy—magnetic and radiofrequency. It also indicated the linkage of chemical with spatial information. However, its potential was not enthusiastically grasped at the moment. In fact, it attracted little attention in the NMR community. Writing in the *Journal of Chemical Education* in July 2004—nine months after Lauterbur received the Nobel Prize in physiology or medicine—Charles G. Fry, director of the Magnetic Resonance Facility at the University of Wisconsin, acknowledges "the embarrassing memory of reading Lauterbur's work in the late 1970s as a young graduate student and deciding (obviously without studying the area in close to enough detail) that the idea of NMRI having practical clinical utility was ridiculous."[8]

Indeed, failure to initially recognize the arrival of a breakthrough concept is not uncommon in science, as in any human endeavor. When in September 1974, a year and a half after Lauterbur's paper was published, Richard Ernst introduced two-dimensional spectroscopy and imaging, a technique readily extensible to three dimensions for the study of biological macromolecules such as proteins, there was no welcoming ovation: "It was little wonder that my...contribution was rated...as premature! Sometimes, however, even prematurely born children survive and excel...Modern times had started without being noticed at that instant."[9] Ernst was awarded the Nobel Prize for chemistry in 1991.

THE PERILS OF PEER REVIEW

The process of peer review by agencies that fund research became dominant after World War II as a safeguard that scientists, not government, determine the pursuit and value of investigations. Previously, wealthy individuals and foundations served as the main source of financial support. But questions regarding the validity of peer review began to arise in the 1970s and became more insistent by the mid-1990s. While peer review is designed for disinterested objectivity, it is not without its flaws. Critics point out that research grant applications force the investigator to work on a problem someone else thinks is important and describe the work in a way that convinces the reviewer that results will be obtained. This is precisely what prevents funded work from being highly preliminary or radical.

Foremost among granting agencies is the federal government's National Institutes of Health (NIH). The NIH invests over $28 billion annually in medical research. More than 80 percent is awarded through competitive grants to in excess of 212,000 researchers at over 2,800 universities, medical schools, and other research institutions in the United States and around the world. Peer review of the 45,000 applications received annually is accomplished by panels of experts called "study sections." In comparison, the National Science Foundation, an independent agency of the U.S. government, has an annual research budget for the biological sciences of $586 million and funds much of the physical sciences research in the United States.

Peer review can institutionalize conflicts of interest and a certain amount of dogmatism. Importantly, the process of peer review was observed as "possibly anti-innovatory" by the prestigious *British Medical Journal* in 1997.[10]

Innovation is praiseworthy, but it is difficult to overcome the inertia of the status quo. Niccolo Machiavelli, the sixteenth-century Italian statesman, knew human nature well: "There is nothing more difficult...than to take the lead in the introduction of a new order of things. Because the innovator has for enemies all those who have done well under the old conditions and lukewarm defenders in those who may do well under the new."

THE MAKING OF A LIFE-LONG ENEMY

Much to his anguish, Damadian's 1971 *Science* article was not cited among Lauterbur's few references. However, Lauterbur did reference an article by federal researchers that referred to Damadian in its very first paragraph. Over time, Lauterbur offered various explanations for this non-inclusion: (1) He was forced to urgently tack on his revision and he had a limit for his bibliography, and (2) Damadian's "controls were inadequate and the publicity overdone."[11] Undoubtedly, however, Lauterbur had another cogent reason for not citing Damadian's work. To obtain a patent, Lauterbur's advance had to be seen as completely novel and independent and not in any way an extension of Damadian's work.

The *Nature* paper brought Lauterbur a lifelong enemy in the person of Damadian, who became convinced that Lauterbur was out to steal his discovery. The view that this was an intentional omission was reinforced later when it became known that Lauterbur's notebook entry on September 2, 1971, credits Damadian's prior discovery. And, Lauterbur, in proposals seeking grant funding, had no hesitation in citing Damadian's contribution. In January 1973, nearly two months before the *Nature* article appeared, Lauterbur filed a federal grant application to the Department of Health, Education, and Welfare, citing Damadian's 1971 *Science* paper and stating that his observation "has provided a new stimulus and goal for NMR studies of tissues."[12] Yet, years passed before Lauterbur openly accredited Damadian in an article in the scientific literature.

Being appropriately cited in the scientific literature is the lifeblood of a researcher's career in terms of status, tenure, and funding. Damadian's experience with Lauterbur's colleague Donald Hollis added to his conviction that he was the target of organized persecution. In a publication in the *Johns Hopkins Medical Journal* in December 1972, Hollis listed Damadian's *Science* article as the ninth reference.[13] In academic circles, burying a reference deeply among others is like "damning with faint praise." About two months after Lauterbur's *Nature* article appeared, Saryan wrote to ask Damadian for material related to points advanced in a recent lecture. In the academic community, this would generally be taken as a compliment, but Saryan, innocently trying to gather information for his Ph.D. thesis, became an unwitting recipient of Damadian's wrath. It was so explosive that Saryan says that he remembers it verbatim thirty years later.[14] "Your group seeks to cheat me of credit for my discovery. You expect me to provide the data as well?" When Hollis,

Saryan's advisor, then called Damadian—who was earnestly working to stake his claim—to see what the problem was, Damadian expressed outrage that Hollis had not properly acknowledged his contribution and demanded that he be cited in the first paragraph, if not the first sentence, of any further publications. Apparently chastened, Hollis complied. Damadian's view was that he, Hollis, and Lauterbur were fiercely competing for funds, and he did not want to be relegated to a secondary position.

Waylon House observed a revealing scene. One day, Lauterbur returned to the lab at Stony Brook with a smile on his face after receiving a phone call from a colleague at the NIH. He had been informed that the NIH had just received a grant application from Damadian and had been urged to submit his own promptly. Lauterbur himself served on review panels at the NIH. "So much for peer review," mused House.[15]

Lauterbur terminated the relationship with Edward D. Welsh, the patent attorney based in Pittsburgh who provided counsel to NMR Specialties, due to a business dispute. In August 1973, Lauterbur turned instead to Research Corporation of America, which acted as an agent in handling intellectual property rights for the State University of New York at Stony Brook, as it did for other universities. Lauterbur made his claim clear that he was "entitled to retain the ownership of a patent under the Patent Policy of the State University of New York." The corporation decided that the expense of pursuing patents would simply not pay off in the end. On February 20, 1974, it informed him that it was unable "to identify a potential market of sufficient size to justify our undertaking the patenting and licensing of this invention." It "did not appear to be sufficient to excite the interest of industry in investing in the development, manufacture and marketing of imaging systems based on this concept" and "its true usefulness and superiority over other already established imaging techniques, such as ultrasonics, radioactive tracers, x-ray, etc., is still unknown."

Lauterbur then requested SUNY to allow him to independently pursue a patent of his imaging method. As he mentions in his Nobel Lecture, university authorities never replied.[16] But the clock ran out. The one-year period for obtaining a patent (after a discovery is recorded, as in a journal article) under the U.S. patent law had elapsed. Lauterbur's quest for intellectual property rights for his method was forced to an end.

Lauterbur had long felt that many are put off by innovation: "Anything new is likely to meet a certain amount of incomprehension at first." Years later, when MRI machines were widespread and Lauterbur had received the Nobel Prize, he offered the ironic understatement that the rejection by the Research Corporation of America and the failure of SUNY administrative authorities to respond to his request to allow him to pursue a patent turned out "not to be a spectacularly good decision."

The timing for funding for Lauterbur's patent application couldn't have been less propitious. In this period, much excitement was being generated over the development of the CT scan for clinical use, enabled by the rapid advances in small computers, and it became obvious that a new revolutionary imaging approach had been born and was taking the world by storm. In April 1972, the EMI Corporation in Britain announced the public debut of its whole-body CT scanner and within three years produced a commercial machine. Radiologists feverishly embraced this advanced imaging method, which combines multiple X-ray images to produce two-dimensional images of the body's organs. By 1977, some 1,130 scanners were in use around the world (more than 700 of them built by EMI), and they soon passed through major developmental phases to become the medical workhorses we are familiar with today. From the very beginning, the scanners were in such demand that EMI salesmen didn't bother to leave the airport when they traveled—hospital representatives had to meet them with cash in hand. Other corporate giants soon came to dominate the marketplace. So welcome was this revolution in diagnostic medicine that as early as 1979, the Nobel Prize in physiology or medicine was shared by Godfrey Hounsfield, EMI's electronic engineer, and Allan Cormack, who devised the mathematics to reconstruct the images—an unusual choice in that the two had never previously met, neither had formally studied physiology or medicine, and neither one had a doctoral degree in any subject.

A BURST OF ACTIVITY

On February 5, 1974, Damadian was granted U.S. patent number 3,789,832 for the human NMR scanner. He had applied in March 1972, one year

3,789,832

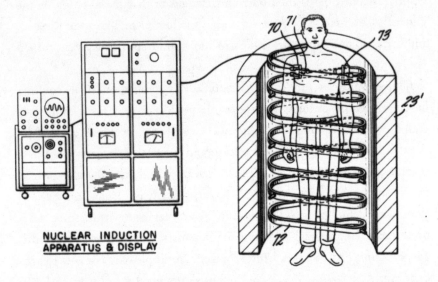

NUCLEAR INDUCTION
APPARATUS & DISPLAY

Figure 9.2   Damadian's patent 3,789,832. An electromagnet shown in cross-section equipped with beam-shaping NMR transmitter coils and radiofrequency receiver probe mounted on a helical track. [RD]

almost to the day before Lauterbur's *Nature* article appeared. Damadian had worked out how he could focus the NMR signal so that the conditions for resonance would be obeyed in only one small area in the body. In this way, it would be essential to take the signal from successive and known points in the body. The patent, for an "Apparatus and Method for Detecting Cancer in Tissues," included the idea of using NMR T1 and T2 relaxation times to "scan" the human body to locate cancerous tissue.

Hollis and Saryan soon thereafter reported elevated relaxation times in other disease states, because the pathological process results in changes of the diseased cells' water content.[17] But they came to what Damadian characterizes as "a foolish conclusion that the technology was worthless." He believed that "it actually opened the door to the early detection of far more diseases than I originally imagined."[18] Damadian remained convinced that relaxation times could be used to detect cancer. His "apparatus" produced no images, and the patent did not precisely describe how such a scan might be done.

Next, Damadian began testing human tissues. In a paper published in the *Proceedings of the National Academy of Sciences* in April 1974, he reported on his studies of surgical specimens of twenty-seven tumors from a wide variety of primary sites.[19] In all cases, the NMR signal was different for tumorous and normal tissues. These he characterized as "chemical fingerprints" and concluded that "this technique should now be considered for use by pathologists as an adjunct to present methods of diagnosing malignancy."

Damadian, meanwhile, knew he was reaching a turning point. Etched in his memory is a declaration made in his presence by Donald Hollis at an international conference sponsored by the Division of Cancer Biology of the NIH in the mid-1970s: "Any more discussion about scanning the human body by NMR is visionary nonsense." This, Damadian laments, "had a damaging effect on my ability to raise funds. My grants were being turned down, they were cancelled."[20] His grant of $225,000 from the National Cancer Institute over a 2^1/$_2$-year period ended in 1976, and applications for further funding to the NIH and to the American Cancer Society were denied. Peter Mansfield at the University of Nottingham shortly published an article in the *British Journal of Radiology* illustrating a cross-sectional NMR image through a human finger—the first, albeit crude, image of a live human subject.[21] This was the first report drawing the attention of radiologists to the potential of clinical applications.

In a burst of activity, multiple laboratories started working and perfecting MRI. In addition to Lauterbur's and Damadian's work, laboratories in Nottingham, Oxford, Aberdeen, and London, as well as the University of California at San Francisco, the Cleveland Clinic, and Massachusetts General Hospital, stood out in this early era. Publications were following each other at a dizzying pace. Advances in MRI were being made, not by the giant commercial vendors who were preoccupied with refining and selling CT units, but by researchers at universities and academic clinical centers. Teams were racing to publish images of ever increasing complexity. Soon the covers and pages of esteemed scientific journals featured images of such objects as a pecan, an onion slice, a lemon slice, a chicken wing, a turkey leg, a mouse, and a rat.[22]

The eventual goal was obvious: to build a whole-body imaging scanner.

Damadian was determined to win that race. He and his team published a crude image of an implanted tumor in the chest wall of a mouse in the December 24, 1976, issue of *Science*.[23] It was obtained as a composite of brightly colored squares after four hours of moving the signal systematically through the live animal. The journal editor put the image on the cover, later explaining it was chosen because of its bright seasonal-looking colors.[24] A few months later, Lauterbur, using his method of zeugmatography, accomplished the same thing and displayed the growth of the tumor over time as well. Surprisingly, he chose not to publish the images for several years, despite pleas by his co-investigators. Today, they can only speculate on the reasons. Perhaps Lauterbur did not want to overshadow his pioneering 1973 *Nature* paper, particularly by appearing to validate Damadian's T1 and T2 concept in tumor imaging.[25]

IMAGING THE WHOLE BODY

With his intense competitiveness and hostility focused on Lauterbur, Damadian was impassioned: "I would have died before I let him beat me." His vision had been the chemical analysis of the human body to obtain quantitative measurements of relaxation times to detect cancer inside the body. But now he clearly realized that only an image of the internal parts of the human body would strengthen his priority claim.

Dating back to an idea he had at the time of his patent, he relied on a focusing technique he called FONAR, an acronym taken from the first and second letters, alternately, of "field focused nuclear magnetic resonance." This would require physically moving the patient across a manufactured "sweet spot," an imaging window up to three millimeters in diameter, for focusing the scan and marking point-by-point relaxation times.

By contrast, refinements of the method of zeugmatography by Lauterbur and then Mansfield would prove to be notably different. Data acquisition from inside the body would be done line by line—that is, by planes—instead of arduously point by point.

Damadian and his devoted postdoc associates, Lawrence Minkoff and Michael Goldsmith, decided that the whole-body imaging unit they wanted could be produced in one way only. They would have to build it, including the magnet, themselves from scratch. Their plans were monumentally ambitious. Their goal was a giant superconducting magnet of 5,000 gauss with a bore (or opening) of 53 inches, large enough to encircle a human being. Superconducting magnets are wound with substances that have no electrical resistance when they're cooled to near absolute zero by liquid helium. They produce very stable fields, yet, once started, they consume no energy at all. It would be, Goldsmith calculated, the ninth largest superconducting magnet in the world. The larger the magnetic field, the better the signal-to-noise ratio to discriminate detail.

Damadian's force of character inspired his associates who gave their all. Joel Stutman, a computer expert, found him "full of fire, full of life. He had a lust for this work. And I have a great deal of respect for someone who will work so hard for an ideal. I had no problem with his personality. I could handle that. Most of the professors at Downstate were afraid of their own shadow and wanted to belong to the same club...I believed in Damadian's dream."[26] Minkoff accepted Damadian's complexity: "He is extremely bright, extremely creative and extremely inventive. He is also very dogmatic."[27] Mike Goldsmith perceived that although Damadian is "a very talented and creative scientist with a flood of ideas, everything gets reduced to a personal fight so that he's combative against his own self-interest. He's more confrontational than he needs to be."[28]

When Damadian presented his plans during a lecture at the Rockefeller Institute in Manhattan, it was clear that conventional-thinking NMR chemists simply could not grasp his vision. Accustomed to spinning a sample in a test tube at one hundred rotations a minute to create a more homogenous magnetic field, someone in the audience jeered: "How fast, Dr. Damadian, are you planning to spin the patient?"[29]

Innovative concepts often threaten the dogma of vested interests. When Louis Pasteur, a chemist, presented his germ theory of infection before the Académie de Médicine in 1878, his revolutionary dictum, which illuminated productive research, practice, and therapeutics, was not greeted with universal acclaim. Some doctors called it "microbial madness" and disdainfully asked Pasteur, "Monsieur, where is your M.D.?"

Bettyann Holzmann Kevles, a science writer, characterizes Damadian's leap in scientific vision aptly: "No one else had the imagination or hubris to...jump to the construction of a whole body machine."[30]

THE DO-IT-YOURSELF MRI SCANNER

Damadian and his team were given an additional room on the floor above their lab and knew they had to break through its thick floor to accommodate the height of their machine below. Goldsmith, a big bear of a man weighing above 300 pounds, spent hours wielding a sledgehammer but to no avail, so they hired a crew with electric jackhammers. They now had a

Figure 9.3   Raymond Damadian, Lawrence Minkoff, and Michael Goldsmith (left to right) and the completed *Indomitable*, presented as the world's first MR scanner.   [RD]

"duplex laboratory," much to the consternation of faculty and, initially, the institution's administration.

Many of the tasks proved Sisyphean at times, but the team persevered with initiative and resourcefulness. It was an enormous leap of faith, and the pressure was unrelenting. Damadian's unshakeable strength of will focused all-out efforts on the goal. His self-confidence could be occasionally unbearable, but those who shared his vision found it deeply reassuring. He took a course in electronics and obtained a computer program from Brookhaven National Laboratory on Long Island, useful in calculating the design of the magnet. He obtained 150,000 feet of surplus superconducting wire from Westinghouse at ten cents on the dollar. Goldsmith, with the help of two graduate students, tightly wound the wire around a metal hoop that was to become the magnet, for weeks on end, six days a week, sixteen hours a day. Minkoff learned how to weld the mile-long lengths of wire from a three-part series in *Popular Science*. Goldsmith found machine tools and motors that were dirt cheap in secondhand stores on Manhattan's Canal Street. Damadian and Minkoff labored over the containers holding the liquid helium and nitrogen, which were necessary to cool the magnet.

The construction of an antenna to pick up the weak radio signals that the unit would elicit from atomic nuclei proved formidable. Goldsmith grimly proceeded by painstaking trial and error on a usable coil, and on the fiftieth try, he reached fourteen inches in diameter, barely large enough to encircle a human chest. With this antenna, which was made of cardboard and copper-foil tape, the machine was complete: a $1^1/_2$-ton, 10-foot-high unit with a 53-inch magnetic opening. Damadian christened it with a name reflecting the team's spirit: "Indomitable." It looked like the entrance to a tunnel. The effort took the men a year and a half, working sixteen to eighteen hours a day, six to seven days a week. Damadian's lab looked like a machine shop, and a visitor would typically imagine a Rube Goldberg creation.

TESTING IT OUT

The magnet proved faulty, woefully short of its goal, delivering only 500 gauss instead of 5,000. Nevertheless, the potential biological effects on a human being subjected to powerful magnetic fields were simply not known. Legend had it that years before, Nobel laureate Edward Purcell had briefly

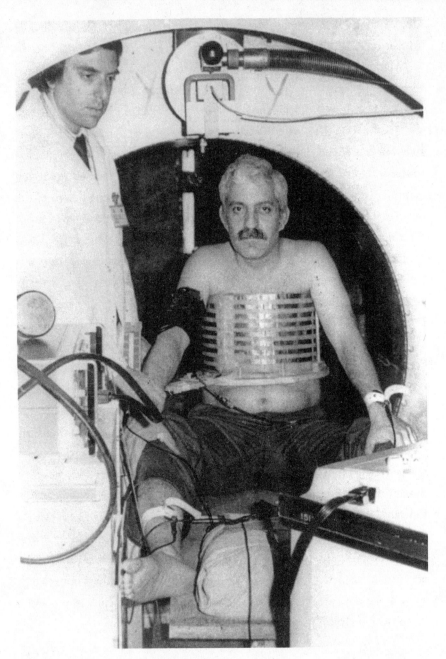

Figure 9.4 First attempt at a human scan. Raymond Damadian sits in *Indomitable*. A blood pressure cuff was affixed to his right arm, an EKG was wired to his chest, and oxygen was kept handy. A cardiologist (standing by at left) was in attendance. Damadian proved oversized for the cardboard vest housing the antenna so that no signal was received from the magnet. A thinner "guinea pig" was needed. [RD]

inserted his head, crowned with a radiofrequency coil, into the 2T field of the Harvard University cyclotron. All he reported was hearing a buzzing and tasting metal from the fillings in his teeth.[31] In the noble tradition of researchers first testing their experiments on themselves, Damadian donned the tight cardboard antenna coil and sat on the beam, with a cardiologist in attendance in case the magnet interfered with his heart. But Damadian was too big, and the antenna failed to work. Nevertheless, while repeatedly exposed to the magnetic field, his blood pressure, pulse rate, and EKGs showed no significant changes.

Critical weeks passed before Damadian and Goldsmith could persuade the slightly built twenty-six-year-old Lawrence Minkoff to volunteer. The machine made measurements at a single, immovable point in space, "the sweet spot" of a dimension about one cubic centimeter.[32] Since the machine was immovable, data could be collected on a point-by-point basis only if the subject shifted his position in relation to the sweet spot.

Minkoff wore the antenna like a vest and sat on a long wooden beam six inches wide within the magnet's circle. The beam was incrementally moved side to side, and Minkoff, sitting on a sliding seat on the beam, could also move backward and forward by increments. At each position, the device scanned a small spot, and those images were assembled. Minkoff had to sit in the machine with his arms raised above his head for four hours and forty-five minutes, bearing a draft through its bore, as Damadian, waiting anxiously to see if his gamble would pay off, assembled a crude image of a two-dimensional section of Minkoff's body from only 106 data points along a plane through the mid-chest.

Damadian wrote in his lab notebook on July 3, 1977:

FANTASTIC SUCCESS!

4:45 a.m. First Human Image

Complete in Amazing Detail

Showing Heart

Lungs

Vertebra

Musculature

Figure 9.5 Lawrence Minkoff sat in *Indomitable* with his arms raised for more than four hours while limited data was obtained as the "sweet spot" changed for a cross-section of his chest at the level of the eighth thoracic vertebra. [RD]

Any human image from the FONAR technique would have been enthusiastically received, but the image of Minkoff's chest was far from "amazing detail." It recalls the famous Abraham Lincoln effect, devised in the early 1960s by Leon Harmon at Bell Laboratories. Harmon produced a picture of Lincoln by digitizing a regular picture into coarse pixels (picture elements). When viewed close-up, the viewer can only focus on the pixels and is unable to see the image of Lincoln. When the viewer moves far away from the photograph or squints, the image blurs somewhat, eliminating the sharp edges, and Lincoln then becomes instantly recognizable. With further computer enhancement, however, the structures of Minkoff's chest become readily distinct.

Critics have pointed out that Damadian relied upon an obscure technology for generating his image. The method could not easily be quantitated

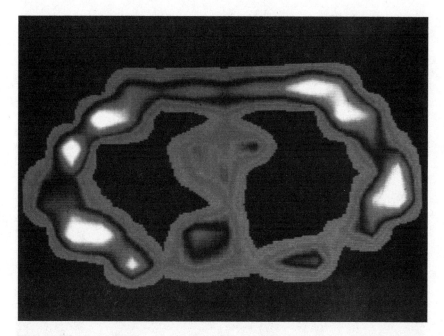

Figure 9.6 First MRI seen of a live human body, further computer-enhanced. Cross-section of Lawrence Minkoff's chest. Top of the image is anterior boundary of the chest wall. Left area is left side of chest. The heart is the principal structure in the middle and the lungs (black cavities) are on either side. More posteriorly on the left, the circular structure corresponds to the descending aorta. In the body wall, the sternum is seen anteriorly and proceeding around the ellipse alteration of light, and dark areas correspond to the intercostal muscles with rib. [RD]

nor reliably reproduced from day to day. It was like settling for a suntan across one line of the body from one beam of sunlight coming through a pinhole. To produce a suntan for the whole body with this technique was beyond imagination.

In an interview with Damadian, I asked him why he persisted with this technique. His response was disarmingly frank. "I was a novice in the field," he declared. "I didn't know from gradients." But he did know one thing: "We showed the world that you can put a human body in a magnet and get an image. Everybody knew then that you could do it. Before that, there was widespread derision."[33]

THE RACE

Damadian had beaten Lauterbur. Along with other scientists, Lauterbur was in the race to produce the first image of the human body using a whole body NMR, but as it turned out, he suffered a stroke of bad luck. A large-bore magnet he had ordered from a private company was delivered without the expected specifications, as a result of the order not having been correctly

Figure 9.7   Paul Lauterbur and his MR scanner with a small aperture, restricting the size of material he could image. [SB]

budgeted for. Its magnet bore was too small to accommodate a human body, and Lauterbur had to wait for a new magnet to arrive.

By then, Damadian was glowing from the telegrams and letters of congratulation from the Nobel laureates Albert Szent-Gyorgyi and Purcell and from University of Nottingham NMR researchers, including E. Raymond Andrew, chairman of the physics department, and Waldo Hinshaw, an American postdoctoral fellow working in Andrew's laboratory. And he received plaudits with what must have been tinged with personal regret from Peter Mansfield.

Mansfield's experience shows that competitiveness and resultant acrimony are not limited to one research group or nationality but rather are inherent components of the race to scientific fame. Mansfield submitted a grant application to the Medical Research Council (MRC) to acquire a whole-body MRI machine, but before sending it off, he first handed it to Raymond Andrew for his comments. To Mansfield's dismay, Andrew shelved it until his own grant application to the MRC—for an intermediate-size imaging machine, one with a bore of only ten centimeters—was granted. Mansfield's application was sent in and approved the following year, but his own human-sized magnet was not delivered until the end of 1977, months after it was expected.[34] If, in his words, "the bloody magnet arrived sooner," he felt he and not Damadian would have gotten the first scan. But he freely expressed his wonder at the achievement: "You have to admire the guy. He set his mind to it and he pulled it off. Anyone who builds a magnet, especially a supercon [a superconducting one], must be an unusual man."[35]

Lauterbur commented later that the FONAR technique was an obvious dead end. The people in his lab referred to it as "Phonymar."

PUBLISH AND PERISH

Now, Damadian felt urgently impelled to publish his achievement of the first image of the human body by NMR in a scientific journal as quickly as possible in order to establish his priority. For this, he purposefully chose *Physiological Chemistry and Physics*, a journal virtually unknown to others. In a recent interview, Michael Goldsmith refers to this as "a house journal," since it was published locally as a forum for researchers outside of the conventional scientific mode who had difficulty having their submissions accepted by the "establishment peer-review system." Damadian served on its editorial board. Unsurprisingly, this move was viewed with great distaste by the NMR community. Damadian essentially molded the concept of "peer review" to his perspective. Goldsmith explains that Damadian, as an outsider to the conventional scientists, felt that his papers were given extra scrutiny and often were required to be corrected or resubmitted, causing undue delay in their publication. "After all," Goldsmith asserts, "the only thing that really matters in the end is whether the experiments can be reproduced and the thoughts have integrative and predictive value."[36]

Recourse to a self-published journal, however, is not without precedent. The great nineteenth-century neurologist Ramón y Cajal was so apprehensive that publication of his enormous output would be delayed, which would threaten his rightful claims to priority, that he financed his own journal and sometimes wrote every article in it.[37]

While Damadian and his team also published in more widely recognized journals, in the five years from 1973 to 1977, they used this journal to publish eighteen articles. Within only a few weeks of submission, his article "NMR in Cancer: XVI. FONAR Image of the Live Human Body" would appear in the journal's September issue.[38]

THE DOWNFALL OF "INDOMITABLE"

Damadian was determined to promote his accomplishment to the world, and in what appears to be a manic state, he arranged a July 20, 1977 press conference to demonstrate "Indomitable," chartering a bus to bring nearly twenty reporters from New York City to his lab at Downstate.

But his press release and presentation were so hyperbolic that it was a debacle in the making.

The press release, approved by Damadian, trumpeted that the machine "pinpoints a chemical characterization of the tumor which will permit determination not only of the type of tumor, but equally, how long it has existed; how wide-spread its metastases; the extent of the malignancy and provide a rational basis for selecting the most appropriate chemical remedy." In a further attempt to establish his priority, the press release held forth the glowing prospect of targeting the radio signals from malignant tissue for radiation "to obliterate it without surgery." The news release contended that Damadian's FONAR technique "is used to discover cancerous tissue in the live patient."

Perhaps carried away by his own vision and zeal, it was embarrassingly obvious to all that his wild claims were premature and grossly inflated. He came across as more showman than scientist, more Wizard of Oz than Wilhelm Roentgen. (Roentgen's discovery of the X-ray earned him the first Nobel Prize in physics, in 1901, which he humbly accepted with two words: "Thank you.")

The press release further announced that reporters would witness the first public application of this technique on a living human. Indeed, attendees did see Damadian's associate, Minkoff, step into the machine and an image of his chest appear on a video screen in color. But one reporter noticed that the date in a corner of the image read July 3 and could therefore not possibly be a "live" image made during the news conference. In Damadian's mind, nevertheless, this was perceived as an iconic image analogous to Roentgen's revolutionary 1895 X-ray picture of his wife's ringed hand.

The next day, Lawrence K. Altman, M.D., a highly regarded medical reporter who had attended Damadian's press conference, wrote a scathing report in the *New York Times* under the headline "NEW YORK RESEARCHER ASSERTS NUCLEAR MAGNETIC TECHNIQUE CAN DETECT CANCER, BUT DOUBTS ARE RAISED."[39] It damaged Damadian's credibility as a scientist exposing as false the claim of having used the technique to detect cancer in a patient, and exposed as untrue Damadian's claim that he used NMR to detect cancer in a patient.

Reading this, Damadian was plummeted into depression. The very fact that he was bypassing the gatekeepers of science by not presenting his group's results in a scientific forum had created not only immediate distrust of his extravagant claims but also a cynical regard of his integrity. He knew he simply could not continue as an academic researcher. Rather than the acclaim he expected for accomplishing a major breakthrough, he received public vilification. Now he knew he would never again qualify for research funds from a granting agency. Rightly or wrongly, many in the academic community, long vexed that he flaunted accepted protocol, now viewed him as something of a snake-oil salesman.[40]

Was the press conference a premeditated act of self-promotion? Bereft of obtaining grants for continuing research, yet convinced of FONAR's prospects to bring about a revolution in "chemical imaging," had he then decided to abandon the academic community as the academic community was abandoning him? Damadian responds: "I was eminently aware from the beginning, very conscious of the financial profits if it worked—my research grants were being cut." He continued, "but at the time, the commercial value wasn't in my mind at all. Just to get credit for the achievement." He knew that Lauterbur was trying to obtain a large enough magnet to fit a human being and that the British were intensifying their competitive goals. "This was the primary concern for the press conference. We wanted the world to know."[41]

A further episode, however, raised doubt concerning Damadian's true motivations. In December 1977, *Popular Science*,[42] a magazine sold at newsstands, featured the cover story, "Damadian's Supermagnet: How He Hopes to Use It to Detect Cancer," and included a detailed interview with Damadian. This only provided further evidence that he was widely promoting his agenda. Ironically, it was this article that first drew the interest of Johnson & Johnson. This mega-company soon formed a new subsidiary, Technicare, for the manufacture and sale of MRI units for imaging patients. They considered putting Damadian in charge of research and development but ultimately settled on Waldo Hinshaw.

Other corporate giants such as Siemens, GE, and Philips, having slowly but surely taken control of the CT market and made huge profits, began to turn their attention to the new opportunities that MRI seemed to offer. Damadian reacted differently from the British groups, which were quite willing to license their patents for royalty-sharing agreements. He saw growing industrial interest as a new and threatening form of competition and was concerned that big business would steal his idea. Their officers came to visit Damadian in his lab, but he dismissed the idea of licensing. "I was very conscious of the fact that large companies bought patents that might be competitive and shelve them and didn't do anything."[43]

Having established priority of publication, he now turned to disseminate his accomplishment in a more widely read, bona fide scientific journal. In May 1978, *Naturwissenschaften*, a German journal of some repute, published a review article on the team's work.[44] Damadian had submitted the article in early January, in response to a direct solicitation from the journal's editor.[45] The image of Minkoff's chest, computer enhanced, was used as a cover illustration. Reflecting an omission on the editor's as well as the author's part, the article was not labeled as a review, and it did not reference Damadian's original paper in *Physiological Chemistry and Physics,* which had been published six months before. Critics charged that Damadian stealthily violated the ethics against duplicate publications.

CHAPTER 10

THE TIPPING POINT

As scientists understand very well, personality has always been an inseparable part of their styles of inquiry and a potent, if unacknowledged, factor in their results. Indeed, no art or popular entertainment is so carefully built as is science upon the individual talents, preferences, and habits of its leaders.

—*Horace Freeland Judson*, The Eighth Day of Creation

Through the trajectory of my professional career in radiology, I was lucky enough to be present for several of the turning points in the development of MRI.

I had first met Raymond Damadian in late 1977 before his departure from the Downstate Medical Center in Brooklyn. I was visiting Joshua Becker, chairman of the Department of Radiology. Damadian excitedly strode into the office to show him Minkoff's image. A primitive display by the FONAR technique overshadowed by the incomparably better resolution being achieved by CT, its clinical promise was then hard to grasp.

I first personally met Paul Lauterbur in July 1978, when I accepted the position of founding chair of the Department of Radiology in the SUNY Stony Brook School of Medicine as well as radiologist-in-chief in the new University Hospital that was scheduled to open in February 1980. My initial task was to plan the hospital department, select and install state-of-the-art imaging, recruit the faculty and technological staff, and initiate a research and residency program. The school had been graduating medical students for several years, but clinical teaching was done at several outside affiliated hospitals.

Lauterbur was a pleasant, highly focused individual with a sardonic wit; his speech was precise, punctuated by frequent thoughtful pauses. While he maintained a full professorship in the Department of Chemistry, I arranged an appointment as research professor of radiology.

PROOF BY BELL PEPPER

Earlier, still restricted to the minute portal of his conventional NMR unit, Lauterbur had produced the first image of a live animal—that of the soft body of a 4-millimeter clam, within its shell, that his daughter found on the beach of the Long Island Sound. Then he was able to secure funding from the National Cancer Institute to build larger magnetic units and to begin assembling a team. Key associates were Ching-Ming Lai, a Ph.D. graduate of Stony Brook's physics department, and, particularly, Waylon House. Both greatly facilitated Lauterbur's speculations with their knowledge of instrumentation and mathematics. It was apparent in Lauterbur's lab, occupying four large rooms in the basement of the graduate chemistry building, that NMR was a burgeoning field of research with fertile applications. But it was one unpublished NMR image in particular that I distinctly remember. It was a detailed image of a cross-sectional slice of a green pepper that clearly delineated its internal structure. To a radiologist used to visualizing internal organs and other structures by X-ray, CT scans, and ultrasonography, the capability to "see inside" without the use of ionizing radiation was an exciting revelation.

Recent NMR literature had displayed images ranging from slices of onions and slices of lemons to sections of a mouse. While these were eye-catching and accepted as preliminary in the evolution of a promising new technology, researchers knew they needed to produce a scan of an intact specimen. Lauterbur's team pondered the possibilities. It was Waylon House who came up with the perfect choice: a whole bell pepper. It not only lent itself to imaging by virtue of being 94 percent water, but it had an internal architecture of septa. The team itself was excited at the details shown in the image, and House was exhilarated. "It's like when you are expending yourself to your utmost to achieve a goal, and you achieve it. Tom Hanks succeeds in starting a fire in the movie *Cast Away*—and beats his chest yelling 'I...I... HAVE...MADE...FIRE.'"[1]

(a)

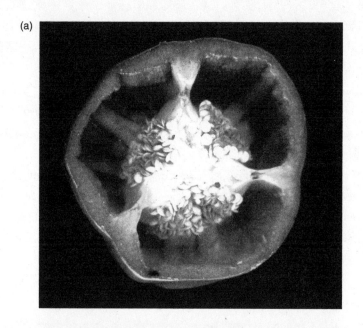

(b)

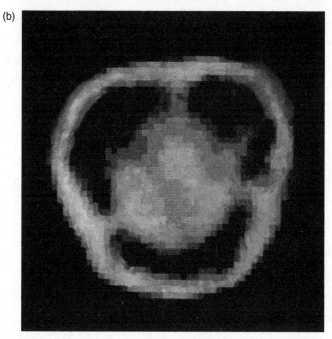

Figure 10.1 Lauterbur's MRI scan (bottom image) of a slice of green pepper (top image). [WH]

Lauterbur first displayed the bell pepper image in the summer of 1976 at the Gordon Conference, held in Tilton, New Hampshire. This was a symposium of researchers involved with the biological applications of NMR. Their reaction was one of astonishment and inspiration. House noted, "You could literally hear an intake of breath; you could hear the collective gasp!"[2] A tipping point had been reached, one that convinced everybody else in the field that the researchers were truly onto something revolutionary. At that moment NMR scientists began to take MRI seriously. House's pride and joy, however, were tempered with disappointment and frustration. Lauterbur chose not to acknowledge the contributions of his team and announced himself as the sole author.[3]

WHAT TOOK THEM SO LONG

From today's vantage point, we may wonder what held back biological investigations for so long since the discovery of NMR. In the foreword to a two-volume textbook on MRI, Edward Purcell maintains that "by 1950, plus or minus a year or two, the basic physics that underlies NMR imaging was for practical purposes completely understood."[4] As early as 1951, Herman Carr, a graduate student in Purcell's lab, initiated the idea of applying field gradients to achieve spatial discrimination—specifically identifying the spatial position of the internal NMR structure of a molecule in a one-dimensional line scan[5]; notably, however, not for image formation. But without any potential use, the one-dimensional map had never been extended to the two-dimensional situation. Lauterbur's practical solution to generate two-dimensional images was to impose successive gradients from different radial directions, and he disclaimed Carr's influence in his conception of NMR imaging.

Another pioneer who produced tantalizing results, Vsevolod Kudravcev, a Latvian electrical engineer at the NIH who had been involved in television technology, devised what appears to be the first MRI device in 1959. Hooking up the output of an NMR machine to a television monitor, he produced a crude image of a quail egg with an embryo inside; his work was based on a Mercator projection, a mathematical model as used for showing a map of the globe on a flat surface. This achievement was mentioned in a progress report the following year by the National Heart Institute. Soon Kudravcev successfully employed static and alternating field gradients to localize the volume of interest. But his supervisor deemed the work a diversion and curiosity and ordered the engineer to abandon his efforts. Kudravcev never published in a journal, though he did include his studies in an occasional lecture. Even today, few researchers know about his work.[6]

Throughout the 1950s and 1960s, there appeared to be no serious scientific justification for investigating biological specimens. Only a few NMR studies of human red blood cells, uterine cervical secretions, saliva, and living muscle were undertaken by a dedicated researcher in Stockholm. As Purcell has noted, the possibility of medically useful NMR images languished for more than twenty years. "What essential ingredients," he asked, "were missing?" Indeed, two were lacking in 1950, he continues: the computer power needed to process the NMR signals and, most important, the idea itself "that a useful interior image was in principle obtainable, and was a goal worth pursuing."[7] CT scanning, which harnessed the computer to traditional technology, had shown radiologists what a technological breakthrough could amount to. MRI, like CT scanning, is inconceivable without modern computer technology.

"What were NMR chemists doing," Damadian rhetorically asks, "for the past twenty years? They were doing chemistry! As an M.D., I had a completely different dimension. I had treated people with disease. I had seen them die." Damadian shakes his head. "You don't forget stuff like that. The need for something like this [NMR study of the intact human body] was preeminent in my head. The need in Lauterbur's head didn't exist, and the need in Mansfield's head didn't exist, until I showed there was the possibility of doing this. I believe one thousand percent if I had not made the discovery and proposed the scanner, I believe that MRI would not exist today." Damadian somberly concludes, "That's what I believe."[8]

After his initial publication, Lauterbur gave numerous talks at universities and society meetings, both in this country and abroad, striving to ignite widespread interest in NMR imaging. He traveled to the United Kingdom frequently and interacted with the Nottingham physicists, specifically Hinshaw, Andrew, and Mansfield, who were convinced that something important would come from their work. In the summer of 1976, on their way to the Gorden Conference in New Hampshire, James M. S. Hutchinson from Aberdeen in Scotland; Waldo Hinshaw and his postdocs; Jay Singer, an electrical engineer from the University of California, Berkeley, who had done pioneering NMR studies on blood flow[9]; and Donald Hollis visited Lauterbur at Stony Brook. Their visit indicates Lauterbur's international reputation; respect among his colleagues led to a closely knit "NMR community" upon which Lauterbur continued to build a position, nurturing the field.

MRI was still an orphan technology, in the mainstream of neither chemistry nor physics. Radiologists, bedazzled by the capabilities of CT, initially thought MRI would offer no competitive advantages. Lauterbur's attempts to arouse interest by industry proved futile. The commercial giants took little or no notice of his invention.

INDUSTRY TAKES OVER

During the next ten years, Lauterbur and his students and colleagues continued to contribute advances in NMR imaging. Virtually all of his publications were in highly specialized journals of basic science.

Those who worked with him in his lab generally found him a caring mentor. Joseph Frank, who worked with Lauterbur as an undergraduate student before going on to medical school, credits him with instilling a passion for science.[10] He is now chief of the Experimental Neuroimaging Laboratory at the NIH. David Kramer, chief scientist at Toshiba's MRI research and development center in San Francisco, collaborated as a graduate student and later as a postdoctoral fellow in Lauterbur's lab in the 1970s and fondly recalls Lauterbur as a dedicated teacher.[11] Today, Ching-Ming Lai is vice president of Analogic Corporation, which is involved in medical imaging equipment, in Peabody, Massachusetts, and Waylon House is an associate professor in the Department of Petroleum Engineering at Texas Tech University in Lubbock.

In the early spring of 1981, as I was serving as a visiting professor in the Department of Radiology at the University of California, San Francisco (UCSF), its pioneering chairman, Alexander Margulis, asked me if I was interested in accompanying him to his MRI research lab. This was in a leased space in an industrial park in South San Francisco. Margulis, an internationally renowned radiologist, had fostered several advances in clinical CT imaging but had been beaten in the race to be first by the Mayo Clinic and the Massachusetts General Hospital. He was determined that UCSF take the lead in MRI. But he had just been informed that his source of funding, the Pfizer Corporation, was leaving the imaging business. Unperturbed, Margulis coolly made a new arrangement with Diasonics for financial support. Typical of the turbulent takeover and merger activity among corporate heavyweights, Diasonics eventually was bought out by Toshiba, the Japanese electronics giant. Within a short time, Margulis's department at UCSF acquired a superconducting magnet and pioneered clinical applications of MRI.

In an editorial for *Radiology*, the premier journal in the field, Margulis stated that radiologists had a new technique that would revolutionize imaging. His statement was rejected as being too optimistic.[12]

Knowledge transfer from the university to industry occurred largely through the import of British-based talent. In the fall of 1978, Technicare, a subsidiary of Johnson and Johnson, hired Waldo Hinshaw from Nottingham; he continued until the spring of 1986, when GE took over Technicare along with its large and innovative engineering team. Paul Bottomley from Nottingham joined the GE Corporate Laboratory in Schenectady in 1980, and his leadership led to a successful effort to make images of the brain. These "stunned" attendees at the Radiological Society of North America meeting in November 1982, the biggest radiology meeting of the year, leading customers to defer purchases of MRI equipment until they could see what GE would offer. Clinical testing of GE's new whole-body MRI units was subsequently done at several medical centers, and physicians at these sites provided considerable feedback and developmental help. Within three years, GE established its place as the market leader.

COMPETITION AND CONFLICT

Clearly, as Lauterbur himself finally acknowledged in 1986, the medical community was first alerted to the clinical opportunities offered by NMR by Damadian's 1971 *Science* report indicating that it had significant potential toward the diagnosis and study of cancer. For twelve years after his breakthrough 1973 *Nature* report, Lauterbur never cited Damadian's original contribution in his numerous publications and presentations. Not a footnote, not a bibliographic reference, not an interesting aside comment. Only when he was awarded the prestigious 1985 Charles F. Kettering Prize by the General Motors Cancer Research Foundation for using magnetic resonance to detect cancer and presented the laureate's lecture published the following year did Lauterbur acknowledge that "the attention of the medical community was first attracted by the report of Damadian that some animal tumors have remarkably long water proton NMR relaxation times."[13] The presentation relied upon images obtained years earlier from his 1977 experiments.

Damadian seethed at the injustice. The detection and treatment of cancer had long been his consuming goal. Furthermore, Lauterbur's colleague, Donald Hollis, had been an outspoken critic of Damadian, initially deriding the non-specificity of his discovery and labeling as "visionary nonsense" his objective of scanning the human body using NMR. While moderating a session that year at the annual meeting of the Society for Magnetic Resonance in Medicine in London, Damadian lost restraint, denounced Lauterbur and the Kettering prize, and started to proclaim his own achievements.

Damadian had felt increasingly marginalized by the NMR and MRI communities for years. In October 1981, an important symposium on NMR imaging, with all the leading imaging groups, was held at the Bowman Gray School of Medicine, Wake Forest University in Winston-Salem, North Carolina. Damadian was invited to present his work, but he astonished the attending scientists by showing a slide of his 1974 patent on detecting cancer and disclosed little about his research efforts.[14] In a photograph of the group of thirty participants, all are smiling for the camera with one conspicuous exception: Damadian.[15] By this time, having established the FONAR Corporation, he was undoubtedly looked down upon as an entrepreneur by the basic scientists.

Damadian's mood was certainly not lightened by Lauterbur receiving the prestigious Albert Lasker Award for Clinical Medicine in 1984. In addition to $15,000, each of the five winners of the Lasker awards that year received a citation and a statuette of the Winged Victory of Samothrace, which symbolizes victory over death and disease. Lauterbur left Stony Brook in 1985 to join the University of Illinois at Urbana-Champaign as a professor

Figure 10.2   Paul Lauterbur around the time he received the Lasker award. [SB]

of chemistry and director of the Biomedical Magnetic Resonance laboratory. Over the following years, time would bear evidence that most of his major contributions were behind him. He died of kidney disease in March 2007 at seventy-seven years old. He had received more than $11 million in funding from the NIH and the National Science Foundation since 1972.

PETER MANSFIELD'S CONTRIBUTIONS

Across the Atlantic, a team from the Department of Physics at the University of Nottingham in the United Kingdom headed by Peter Mansfield had been independently developing a way of using magnetic field gradients to obtain spatial information of a solid sample inside an NMR machine. His first publication on the principle of NMR imaging of solids to the level of atomic structure followed Lauterbur's *Nature* article in the same year. When Mansfield presented his paper at a physics conference in Krakow, Poland, in September 1973, someone in the audience asked if he knew of similar work that had been carried out and published a few months earlier by Professor Paul Lauterbur in the United States. Startled, Mansfield was, in his own words, "completely unaware."[16] He thought he was telling the world for the first time about NMR imaging. Indeed, this episode recalls the simultaneous discovery of NMR by Felix Bloch at Stanford and Edward Purcell at Harvard, two independent groups, unknown to each other, working on opposite coasts of the United States. They published their results in 1946, in consecutive issues of the journal *Physical Review*. As I mentioned in chapter 4, science is replete with instances of simultaneous independent advances, influenced by a zeitgeist, or spirit of the times. When ideas are "ready," they seem to spring spontaneously and independently from many sources.

The person raising the matter was John Waugh of MIT; he had been engaged in a stormy and acrimonious rivalry with Mansfield for several years. Taking a cue from Lauterbur's paper, Mansfield turned his attention from the physics of solids to the biological framework of largely liquid materials. Within a few months, he showed how scientists could isolate a thin layer of a sample, enabling only that "slice" to resonate. In this way, researchers learned how to use NMR to make three-dimensional images. He provided the final leap in overcoming the problems that prevented usable data from being acquired on living, moving (breathing) subjects in 1977 when he published a method called echo-planar imaging. Full three-dimensional images could now be acquired up to 10,000 times faster, allowing detailed studies of all portions of the human anatomy. His work also enabled NMR imaging to capture moving or flowing images, like a heart pumping blood.

MRI quickly advanced into new realms: real-time moving images of heart function, for example, were now possible. And Mansfield's team, supported by a grant from the Medical Research Council, produced a crude but promising line-scanning image through a live human abdomen (that of Mansfield himself) in only forty minutes; results were published in the November 1978 *British Journal of Radiology*.[17]

Eleven months earlier, another team, headed by Waldo Hinshaw, had published an NMR image of the human wrist in *Nature*.[18] Generations of medical students, confronted with the Latin names of the wrist bones, relied upon the mnemonic "Never Lower Tillie's Pants, Good Man, Lest Mother Come Home" for the navicular, lunate, triquetrium, pisiform, greater multangular, lesser multangular, capitate, and hamate. Many of these, as well as soft tissue structures such as tendons and ligaments, were clearly displayed.

These two reports encouraged manufacturers and directed the clinical imaging of MRI to radiologists.

It was Mansfield's advances that allowed MRI to become practical as a clinical technique. He was knighted in 1993 and shared the Nobel Prize with Lauterbur ten years later. The University of Nottingham did file patents that made Mansfield wealthy enough to donate a new MRI center to the university.

FONAR

Nine months after achieving Minkoff's chest image, Damadian left Downstate along with a few of his associates, including Minkoff and Goldsmith. With venture capital he solicited, Damadian founded the FONAR Corporation in Melville, Long Island. He still heads this company, which produces and markets commercial MRI machines. Damadian retained 51 percent controlling shares, Minkoff and Goldsmith each received 7 percent, and Stutman 3.5 percent, which they appreciated as a generous arrangement.[19]

Damadian's challenge was to channel his energy in his new role as an entrepreneur. He continues to maintain that his long-term goal of conquering cancer has remained the same. In 1980, the company introduced the first commercial MRI scanner at the annual conventions of the American Roentgen Ray Society and the Radiologic Society of North America. The usefulness of the technique to pathologists had long been abandoned, and he was now clearly promoting its imaging capabilities to radiologists. His scanner was based on an open permanent magnet, an iron core with brick magnets, and an improved field-focusing technique, to generate slice images quickly with better detail. Permanent, rather than superconducting, magnets were chosen because they had lower running costs and were less troublesome for hospitals to bother with. Furthermore, an open magnet allowed the patient to extend his or her arms and did not bring about the sense of claustrophobia often encountered in a conventional cylindrical unit. FONAR went public in June 1981 and the following year introduced a much more powerful permanent magnet scanner.

Major vendors promptly began to market superconducting MRI units with crisper images. Commercial units were produced by a host of companies.

Between 1977 and the early 1980s, the machines were marketed principally to radiologists. Nuclear medicine specialists had a faint hope that the term NMR might foretell "no more radiologists," but marketers dropped the word "nuclear" from the name for fear it would invoke images of lethal radiation and tagged on "imaging" to solidify the bond with radiologists.

As part of a promotional campaign for the FONAR Corporation, Damadian mailed thousands of copies of a book titled *A Machine Called Indomitable* to radiologists.[20] Written by Sonny Kleinfield in a generally hagiographic manner, the book detailed the Herculean effort in building the machine, but it was in laying bare aspects of Damadian's character that the expected response backfired. A review in the *New York Times* opened with, "It is a tribute to the [author's] honesty that the reader may feel admiration for the scientific accomplishment described and less than admiration for the scientist who accomplished it."[21]

CHAPTER 11

OBSESSION

Our religion, our moral fabric, our very basis of life are centered round the idea of reward. It is not abnormal therefore that the research man should desire the kudos of his own work and his own idea. If this is taken away from him, the greatest stimulant for work is withdrawn.

—Sir Frederick Banting, Nobel laureate,
discoverer of insulin

Raymond Damadian, feeling that his fundamental discovery that the relaxation times in magnetic resonance can discriminate between different tissues in the body had long been unethically appropriated by others, now embraced the challenge of a relentless legal battle with corporate giants that would consume him for the next fifteen years.

In 1982, FONAR filed suit against Technicare, which had rapidly acquired a major market share. It produced about 100 of the estimated 250 MRI machines then in use, at a price of $1 to $2 million each. MRI was increasingly finding widespread use, accelerated by CT having paved the way for an imaging system that did not rely on radiation. Sales of units were further boosted when the federal government ruled that doctors and hospitals could be reimbursed for using the device to treat Medicare patients. A jury decided in FONAR's favor, but four years later, after appeal, a federal judge overturned the jury's finding on a technicality. Technicare was shortly bought out by General Electric.

Damadian abandoned his point-by-point FONAR technique and adopted Lauterbur's method, but that approach was quickly superseded by further improvements in the MRI industry. Struggling to maintain the FONAR Corporation's viability, Damadian opened his own MRI scanning centers across the country to serve doctors and hospitals. He hoped to maintain FONAR'S tiny market share with its open machine, but in 1988, the Hitachi Corporation came out with an identical unit.

And then a circumstance occurred that radically changed the company's prospects. Honeywell Inc. successfully sued the Minolta Camera Company for patent infringement on auto-focus cameras for more than $200 million in back royalties. In September 1992, FONAR retained on a contingent fee basis the Minneapolis law firm that won the case. Even though Damadian's original patent had expired in 1989, the lawyers realized they could sue not only for royalties but also for lost profits. MRI had become a $2 billion industry. FONAR filed suit against Hitachi, GE, and other multinational companies for infringement of the original patent to detect cancer as well as a patent for a technique called multi-angle-oblique (MAO) that did not expire until 2006.

The argument for infringement of the first patent invoked a legal principle called the "doctrine of equivalents"—that is, unauthorized use substantially equivalent in function or operation to that literally claimed in the patent. Damadian had patented the original application of relaxation times to characterize and distinguish the body's tissues. Fat, for instance, has different relaxation times as compared to water. As a result, relaxation times serve as very useful parameters for the construction of MRI images.

The importance of the discovery of the diversity of tissue relaxation was stressed by no less an authority than Felix Wehrli, MRI scientist and former manager of the NMR Applications Division at General Electric. In a June 1992 article in *Physics Today*, he affirmed, "This difference provides the basis for image contrast between normal and pathological tissues."[1] The MAO patent covered a technique to obtain multiple image slices of a patient's body at different angles in a single scan. Damadian argued that, "It's a key innovation we invented, particularly for evaluation of spinal disks." GE claimed that "We've been developing these concepts from the beginning" and had obtained nearly 250 patents in the field of MRI.

By 1996, one by one, companies settled: Hitachi, Siemens Medical Systems Inc., Philips Electronics.

And then, on June 30, 1997, after a five-year court battle, the U.S. Court of Appeals for the Federal Circuit upheld a lower-court ruling that GE had violated Damadian's patents and ordered the mega-Goliath to pay him $128.7 million. The judge declared that GE's *use* of T1 and T2 MR images was an insubstantial difference from Damadian's proposal in his original patent. Within a few months, the Supreme Court denied GE's petition for appeal. All MRI methods rely on tissue differences of proton relaxation as discovered by Damadian in 1971. In this way, the U.S. justice system, up to the Supreme Court, under the rules of evidence, clearly affirmed Damadian's priority. Two days later, FONAR had the money in its bank account. *FONAR Corporation v. General Electric Company* has become a classic case in jurisprudence pertaining to patent violations of scientific claims.[2]

By court order, the amount of settlement in each case must remain undisclosed, but when questioned, Damadian allows that the total of these "approaches" the GE sum.[3] The judgment money has all been put back into FONAR for research and development purposes.

INTELLECTUAL PROPERTY AND PATENTS

From the earliest days of the nation, the intellectual property rights for an invention have been guaranteed to the inventor. The U.S. Constitution gives Congress the power to enact laws relating to patents "to promote the progress of science and useful arts, by securing for limited times to authors and inventors the exclusive right to their respective writings and discoveries." The purpose of the patent system is to stimulate innovation. With a patent, the inventor is guaranteed the right to exclude others from making, using, offering for sale, or selling the invention, or importing the invention into the United States.

To be granted a patent, an idea must be novel, capable of being made, useful, and non-obvious to someone with a good knowledge of the area involved. How to define invention more precisely beyond parameters of novelty and usefulness has eluded lawyers since Thomas Jefferson struggled with the concept over two centuries ago. The Doctrine of Non-obviousness—that a patentable invention must be obscure to a person of ordinary skill in the field—became law in the Patent Act of 1952. Justice William O. Douglas, writing on this subject, stated that an invention, "however useful it may be, must reveal the flash of creative genius, not merely the skill of the calling." Critics allege that since most advances follow from a line of antecedents, this is a somewhat romantic notion, but an individual with a breakthrough intuitive inspiration can often be identified.

Whereas invention involves making something that is completely new, discovery is finding, by chance or diligent search, something that already exists. There is, however, a fascinating relationship between the two processes because they feed on each other, with one invention or discovery leading to another. However, the discovery of a natural law, physical phenomena, or abstract idea is not patentable. A scientist cannot patent a natural principle he or she reveals—only some new application of it. Indeed, patents were not always of paramount importance in research; many academic scientists once believed that knowledge should not be treated as proprietary. Röentgen was not interested in patenting his discovery of the X-ray and refused various financial inducements to do so. After Jonas Salk invented a polio vaccine, the journalist Edward R. Murrow asked him who owned the patent. "There is no patent," Dr. Salk famously replied. "Could you patent the sun?"

The picture is far different today. Congress undertook major new initiatives to use patents to spur industrial growth, among them the 1980 Bayh-Dole Act, which fostered university-industry collaborations by making it possible for universities to patent the results of federally funded research. University-based scientists share in royalties generated by licenses.

Today a patent provides a legal monopoly for a term of twenty years from the date the application is filed. It generally takes two and half to three years for the patent to be issued. The U.S. Patent Office is an agency of the Department of Commerce and employs thousands of patent examiners. It has issued over seven million patents. To understand the significance of Damadian's courtroom victory, it's useful to know that only 1.5 percent of patents become the subject of lawsuits; only 0.1 percent are litigated to trial; and courts hold about 45 percent of all litigated patents invalid.[4]

Damadian has received over sixty patents (some co-invented) for improvements to his MR scanner. Among his innovations are a unit that gives unrestricted patient access, allowing surgeons to operate while patients are being scanned, and the Upright MRI™—the only MR unit capable of imaging patients while they are standing or sitting. Damadian is an ardent advocate for patent protection and believes it encourages the independent inventor.

PATENT TROLLS

In the past several years, individuals or companies known by the derogatory name of "patent trolls" have exploited vulnerabilities in the existing laws. Patent owners need not commercialize the invention to enforce their patents. Patent trolls acquire patents not to manufacture and distribute inventions but to license them, under threat of litigation if necessary, to companies that have unwittingly crossed the patent's boundaries.

Patent trolls may anticipate the direction of a developing field, acquire patents directly or buy unused patents with potential from others, and wait for the proper time to legally press their rights upon an unsuspecting company. Their pattern is to extort hefty fees in licensing negotiations as well as huge settlements from companies they have accused of infringement.

Even as early as 1941, the U.S. Supreme Court aptly spotlighted "a class of speculative schemers who make it their business to watch the advancing wave of improvement, and gather its foam in the form of patent monopolies."[5] The technique was perfected by Jerome Lemelson, an engineer and inventor, in the 1980s and particularly the 1990s. "Lemelson didn't patent inventions," one of his former lawyers quipped, "he invented patents."[6] Described as a "workaholic," he spent twelve to fourteen hours a day writing up his ideas. Over a period of forty years, he received an average of one patent a month in a myriad of fields from medical and industrial technologies to gadgets and toys. With an initial $2 million settlement from Sony and a $5 million license negotiated with IBM, he and his attorney threatened lawsuits based on his over 600 patents, many guessing the path that developments would take. They won huge settlements against Japanese, European, and American automakers and Japanese electronics manufacturers. Other companies targeted included Siemens, Motorola, Kodak, and Apple.

Lemelson perfected the use of what came to be known as "submarine patents" to negotiate profitable licenses. By filing a succession of continuation applications delaying the final approval of patents, they may be first published and granted only long after the original application was filed, sometimes over several decades. They then emerge, taking the whole market by surprise, to extract huge settlements from major corporations in a variety of industries. By the time of his death in 1997, Lemelson had reportedly netted the astronomical sum of $1.3 billion. In 1995, the Patent Office changed the period of patent protection from seventeen years from the date of issue to twenty years from the date of filing. No longer was there an advantage in using submarine patents. Further, in a 2004 landmark court case, the continuing claims of Lemelson's estate were determined unenforceable, based on the doctrine of *laches* (negligence or undue delay in asserting a legal right or privilege).

Lemelson is a highly controversial figure, hailed as a champion and a staunch advocate by the community of independent investors, excoriated by some of the companies he sued.

THE FIGHT GOES ON

Emotions based on institutional and national loyalties still run high. The Downstate Medical Center occasionally issues protests of injustice at the Nobel committee's decision to exclude Damadian from the prize. Stony Brook scientists shake their heads in dismay at Damadian's continuing lamentations. In 1999, he was invited to give a talk at an annual lectureship at Stony Brook, and scientists in attendance still recoil at the memory of his "egomercial" and rants against Lauterbur. Just as Damadian's obsession persisted, so did Lauterbur's continued efforts to erase Damadian's contribution from memory. In a 1996 encyclopedic history of NMR, Lauterbur included a notebook entry written on the evening of his epiphany twenty-five years earlier. No doubt purposely omitted was the citation of Damadian's original discovery.[7]

It was, after all, Damadian—a physician with the vision of a zealot— who conceived of placing a human in a giant magnetic resonance unit to derive data of medical use. In another context, speaking of using NMR for imaging, Lauterbur contended, "No one had thought that there was a way to do it. If it doesn't seem possible, nothing much gets done."

Donald Hollis's 1987 self-published book, *Abusing Cancer Science: The Truth About NMR and Cancer,* written as an avenging angel deriding Damadian's efforts, has become something of an underground classic at Stony Brook. Yet, C. N. Yang, the 1957 Nobel laureate in physics and the Albert Einstein professor emeritus at Stony Brook, has testified that the prize should have been shared by Lauterbur and Damadian.[8]

Damadian understandably must have taken comfort from the numerous awards he had received. In the late 1980s, President Reagan jointly awarded the National Medal of Technology to Damadian and Lauterbur "for their independent contributions in conceiving and developing the application of magnetic resonance technology to medical applications including whole-body scanning and diagnostic imaging." At the ceremony, Lauterbur extended his hand to Damadian, who chose to dismiss the offer by turning his back. The following year, Damadian was inducted into the National Inventor's Hall of Fame, which was established by the U.S. Patent Office, as the sole inventor of MRI. Previous recipients included Thomas Edison, Alexander Graham Bell, and the Wright brothers. "Indomitable" was acquired by the Smithsonian National Museum of American History in 1986 and is now on display at the National Inventor's Hall of Fame in Akron, Ohio.

In 2001, the Lemelson-MIT Prize Program, endowed by "patent troll" Jerome Lemelson in 1994, bestowed its $100,000 Lifetime Achievement Award (the "Oscar for Inventors") on Damadian as "the man who invented the MRI scanner." And on September 23, 2003, two weeks before the announcement of the year's Nobel Prizes, making the disavowal of Damadian all the more painful, he was honored with the Innovation Award in Bioscience from *The Economist,* for "using the principle of nuclear magnetic resonance... leading to the creation of the MRI industry."

A few days after the Nobel Prize ceremony in Stockholm, in December 2003, a profile of Damadian appeared in the *New York Times Magazine.*[9] The subtitle of the magazine's profile—"Scanscam?"—was pointedly injudicious. In his incensed reply to the interviewer's statement that "some scientists have said that although you built an early scanner for medical applications,

MRI as we know it today rests on Lauterbur's advances in making a discrete image," Damadian likens himself to the Wright brothers.

One hundred years earlier, almost to the day, on a windy, chilly morning, starting from a North Carolina sand dune called Kill Devil Hill, Orville flew the brothers' 600-pound plane through the air in an epochal journey lasting twelve seconds and covering 120 feet. Wilbur and Orville's first airplane did not require a government grant, for they constructed the wings and fuselage themselves, with tools from their bicycle business, and had their resourceful bicycle mechanic design and hand-build the engine. Their spirit, too, undoubtedly in Damadian's mind, could be called indomitable. Damadian notes that airplanes now all use ailerons, Glenn Curtiss's invention, to control banking in turns. Yet the Wrights get the credit for the first effective flight of a powered, heavier-than-air craft. Damadian likened himself to the Wrights and Lauterbur to Curtiss. Mansfield was not mentioned.

On a recent sunny spring afternoon in his office on the sixth floor of the Graduate Chemistry building at Stony Brook, with windows overlooking the university's green athletic fields, Professor Arnold Wishnia heatedly offered his reaction to this:

"Damadian notes that 'the Wright brothers had a very primitive technique for controlling flight, yet they, not Glenn Curtiss, get the credit for an effective flight of a powered, heavier-than-air craft, because they did it first.'" Damadian, Wishnia asserts, has it backward. "Well, in fact, the Wrights analyzed all the main problems correctly, and imaginatively chose, adapted and improved the available technology to solve them. They flew. All airplanes today follow their essential plan: proper lift from fixed wings, horizontal stabilizers and elevators, a rear rudder, an engine of proper weight and power. For the analogy to be truthful, Lauterbur should be likened to the Wrights and Mansfield to Curtiss. Damadian's 'Indomitable' was a contraption that flapped itself to pieces as it bounced along the ground."[10]

And Leon Saryan, now the technical director of the industrial toxicology laboratory at West Allis Memorial Hospital in Wisconsin, has been the object of remonstrations for his unqualified declaration that Damadian did not deserve the Nobel Prize. An American Armenian, like Damadian, Saryan has been castigated in Armenian press as a traitor.[11]

A CREATIONIST

The Internet is buzzing with rumors that another consideration may have influenced Stockholm's decision to spurn Damadian. In the early 1960s, Damadian attended a Billy Graham crusade rally at Madison Square Garden and became a born-again Christian. In recent years, he has come out as a "creation scientist," convinced of the scientific truth of Genesis. He believes the world was created by God in six days about 6,000 years ago and rejects Darwinian evolution. "The idea that all modern life oozed out of the bottom of the sea," he declares, "is preposterous." He is convinced that the Bible is the reason for the advancement of science and the blessings of Western civilization.

Michael Ruse, professor of the philosophy of biology at Florida State University, speculates that Damadian's exclusion was motivated by his support of creation science. Writing for the Metanexus Institute, he has said, "I cringe at the thought that Raymond Damadian was refused his just honor because of his religious beliefs."[12] When asked about this, Damadian says he prefers not to think it was a factor. He was never directly confronted over the issue, "but maybe, people are too polite to ask." By contrast—and this may be in no way relevant to the Nobel committee's decision—one of Lauterbur's later areas of expertise was the chemical origin of life. He investigated possible transitions from chemistry to biology through analysis of molecular imprints.

Dr. Francis Collins, the head of the NIH, and William Phillips, the 1997 Nobel laureate in physics, both trained scientists, can comfortably accommodate science as a matter of faith. "I find my appreciation of science is greatly enriched by religion," Collins has said. "When I discover something about the human genome, I experience a sense of awe at the mystery of life, and say to myself, 'Wow, only God knew before.' It is a profoundly beautiful and moving sensation, which helps me appreciate God and makes science even more rewarding for me." Reconciling faith and science in the tenet of evolution, Collins accepts that "a creator God set the process" of evolution in motion.[13] Phillips speaks of the joy it brings him to know that God cares for him, and he doubts that any scientific experiment will ever support or destroy his belief in a personal God.[14] "I do believe, more *because* of science than in spite of it, but ultimately just because I believe. As the author of Hebrews put it: 'faith is the substance of things hoped for, the evidence of things not seen.' "[15]

Armenians have a particular kinship with cosmic events portrayed in the Old Testament. Creationists have postulated the likely size and shape of Noah's Ark. Others have conjectured that the flood began with a massive volcanic eruption of Mount Etna in Sicily that produced a tsunami of epic proportions across the Mediterranean Sea. If one stands in Yerevan, Armenia's capital, and looks southward, Mount Ararat is conspicuous on the horizon. It is here, the Bible relates, that Noah's Ark settled as the flood waters receded. A piece of wood claimed to be from Noah's Ark is preserved today as a precious relic in the capital's main church. Requests to submit a small sample of the wood to carbon dating to determine its true age are politely declined on the basis that the legend's acceptance relies upon faith. It is a fact that 60 percent of Americans literally believe in the story of Noah and the flood. Armenians see themselves as the direct descendants of Noah and are proud that the remains of Apostle John are entrusted to them in the Armenian quarter of Jerusalem.

Now, Damadian expresses a radical change in his view of the Nobel Prize. "I passionately wanted to get it. It's ironic but I'm grateful that the good Lord spared me the dishonor of the Nobel. In all honesty, actually if I had been offered it in 2003, I would have accepted it with great jubilation. But I don't see it that way today—I just don't. The gratification in the big picture was misdirected. What I came to realize, and I'm grateful for the experience, is that man's greatest weakness is that he lives for the approval and acknowledgement of mankind. The proper objective of such aspirations is for the glorification of the Lord and not the individual."

I asked him why he gets up every day and comes to work. Is there not some element of human ego? Without hesitation, he answered, "That's not where I want to go. It's not my goal. I want and need a successful financial enterprise to get where I want to go. To make a successful attack against cancer. I had to make detours in order to maintain financial stability of the company. Now that I have the technology—for example, the open surgical MRI for ablation of tumors—I can resume my goal."[16]

As I rose at the end of my last interview with Damadian, to shake his hand in thanks, I noticed his tie was designed with the word "Genesis." Ever the evangelist, he thrust two books on creation science by a leading advocate in the field into my hands as I was departing.

WHO WAS IN THE RIGHT?

Raymond Damadian's typically defiant stance toward being excluded from the Nobel Prize earned him neither sympathy nor support at the time. An immediate response was made by Horace Freeland Judson, a science journalist who had written *The Eighth Day of Creation*, a major book on the history of molecular biology. Judson fired off a peevish op-ed piece in the *New York Times* headlined, "No Nobel Prize for Whining," chiding Damadian for his "noisy complaining."[17] Judson pointed out that several others in the past, deserving but unrewarded, suffered in silence. Judson's condemnation was a gratuitous slap that denied, in effect, a scientist's legitimate striving for the claim of priority.

Damadian's newspaper campaign cost a total of $1,200,000. He was determined to validate himself. Based on his central patent on MRI technology, awarded in 1974 and affirmed by the Supreme Court in 1997, Damadian and the FONAR Corporation had won hundreds of millions of dollars in patent infringement penalties from the international MRI vendors. But he was unable to win the Nobel.

To those in the NMR community, Damadian's campaign was viewed as an extension of his long-standing abrasive behavior and outrageous claims. But the Nobel committee rewards scientific achievement, not good manners. To some, the question remained of how the crucial steps in invention are valued, and opinions at the time varied. "Damadian deserved to be considered for the Nobel," said Stephen Thomas, himself an MRI experimenter at the University of Cincinnati Medical Center and a past president of the American Association of Physicists in Medicine. His judgment to a reporter then was that Damadian's 1971 paper in *Science* "laid the groundwork for using magnetic resonance parameters in diagnosis."[18] John Gore, who directed the Institute of Imaging Science at Vanderbilt University, also thought it would have been fairer to include Damadian in the prize. "There is a case to be made," he stated, "that he was a visionary and proposed the idea of scanning through the human body."[19]

Some, however, harbored deep resentment against Damadian. Gary Fullerton, an MRI specialist at the University of Texas Health Sciences Center in San Antonio and founding editor of *The Journal of Magnetic Resonance Imaging*, was unguarded in his comment: "As I look back, if Dr. Damadian hadn't been so vitriolic in his attacks on other people, I would have argued that he deserved to be included [in the prize]. He created this situation himself." Fullerton recalled Damadian's diatribe against Lauterbur at a session of the 1985 London meeting with several hundred scientists in attendance. "People were absolutely shocked that he did that. That sort of behavior really doesn't set well with scientists."[20] To many, Damadian's actions had cast him in the role of Pogo, who said, "We have met the enemy and he is us."

In contrast, many researchers accept Lauterbur's invention as the fountainhead of MRI. Ian Young, an electrical engineer at Imperial College in London and a former member of the EMI/Hammersmith collaboration, maintained that the "key invention" that everybody uses in MRI today "is the gradient field." It was Lauterbur, most felt, who made imaging a reality. Yet, the use of the word "invention" for Lauterbur's innovative contribution raises, in the minds of some Damadian supporters, a critical issue. Alfred Nobel's 1895 will clearly distinguished between discovery and invention as they regard the three major scientific fields of physics, chemistry, and physiology or medicine: "One part to the person who shall have made the most important discovery or invention within the field of physics, one part to the person who shall have made the most important chemical discovery or improvement, one part to the person who shall have made the most important discovery within the domain of physiology or medicine." It cannot escape notice that the award in medicine requires discovery and was not meant to honor invention. Lauterbur performed another critical function: He was a tireless evangelist in advocating MRI. According to Waldo Hinshaw, originally a researcher at the University of Nottingham who then worked for an MRI manufacturer for several years, Lauterbur "nurtured the field until it attained a life of its own. He visited our lab several times while I was working in England. He paid for me to visit his lab in Stony Brook. He was open with his ideas and full of encouragement. He made sure, through these efforts, that the early developers stayed in touch and were friends."[21] For the most part, Damadian was widely recognized outside of the academic community as an entrepreneur and only begrudgingly as a progenitor. Lauterbur was perceived by most as a nurturer.

Their contrasting personalities determined the scene. Damadian, convinced that Lauterbur's contribution was only a technical refinement, rebelled against the constraints imposed by the "insiders" and was viewed as resorting to secrecy, dubious self-promotion, and commercial self-interest. Lauterbur, on the other hand, denied in his pursuit of both a patent for his invention and consequently support by the major commercial vendors, burnished his image as nurturing the field.

In his Nobel lecture, Lauterbur very precisely made the distinction, starting with his failed efforts to patent his breakthrough:

> So I abandoned that idea and decided instead to encourage others to pursue this new technology, inviting everyone interested to visit my laboratory to observe our efforts and learn from us. People did come, from industry, academia, and government laboratories, foreign and domestic, and I began supplying a bibliography of such work to all and helping to organize meetings on the subject to compare our methods and results...
>
> As the depth and breadth of application grew, both large and small companies began to see opportunities, and within less than ten years commercial instruments began to come to market, large enough to hold a human being and to support true clinical research. Competitive pressures among physicians, industrial interest, and multiplying applications and techniques began to generate the explosive growth that was to characterize the past twenty years... I and my group continued to make contributions though this period as well, some of them significant.[22]

While facts can be established, the determination of who merits priority in discovery requires judgment. Damadian's and Lauterbur's roles in the story cogently illustrate one of the observations of Francis Galton: in history the driving force may not lie with the first discoverer of a new scientific fact, but rather with the individual who was the first to persuade the world of the importance of a particular discovery.

CHAPTER 12

PICKING THE WINNER

The scientific community often finds it difficult to accept new ideas...The greatest and most harmful source of resistance from scientists to scientific discovery comes precisely from those peers whose mission is to preserve the quality of scientific work.

—*Juan Miguel Campanario*

Every development and every discovery in science can be traced back to antecedents that provided basic theory or applicable technology ready to be exploited. Howard Florey detailed the series of incremental step-by-step contributions over decades to the field of antibiotics in his massive review *Antibiosis*. In this way, he acknowledged the heavy indebtedness to the legacy of knowledge bequeathed by predecessors. During a speech to the Royal Academy, Florey said: "Science is rarely advanced by what is known in current jargon as a 'break-through,' rather does our increasing knowledge depend on the activity of thousands of our colleagues throughout the world who add small points to what will eventually become a splendid picture much in the same way that the Pointillists built up their extremely beautiful canvasses."[1]

But giant steps were taken, first by Paul Ehrlich in 1909–1910 with the discovery of an arsenical compound, Salvarsan, effective against the spirochete that causes syphilis. This was followed by Gerhard Domagk in the early 1930s with the discovery of the sulfa drugs, useful against streptococcal and other infections. In 1940, Howard Florey and his chemist Ernst Chain developed penicillin, pursuing Alexander Fleming's chance observation eleven years earlier of the anti-bacterial action of a mold. And in 1943–1944, Selman Waksman and his graduate student Albert Schatz uncovered a soil microbe effective against diseases that had been unaffected by penicillin, particularly tuberculosis.[2]

MRI was made possible by contributions from numerous visionary scientists—including a handful of Nobel Prize winners—over a period of more than fifty years. Researchers incorporated bits and pieces from many different disciplines, such as chemistry, mathematics, engineering, computer science, medicine, and, of course, physics. NMR can be said to have taken origin from the concept of electron spin and nuclear spin as a quantum mechanical entity; as NMR phenomena and theory were better understood, the technique could be applied in many fields. Major improvements in instrumentation and technology have fostered ever-expanding applications. Building on Raymond Damadian's discovery of tissue differences, Paul Lauterbur brought about medical imaging by coupling computer analysis to the new data inherent in applying gradient fields.

Thomas Kuhn, a widely cited philosopher of science, distinguishes between two general types of science: "normal science" and "revolutionary science."[3] In "normal science" there is conventional research with its trial-and-error improvement and incremental extrapolation of already existing paradigms. This corresponds to Florey's concept of scientific pointillism. But Israeli philosophers of science Aharon Kantorovich and Yuval Ne'eman take a dim view of this: "Since the scientist is in general 'imprisoned' within the prevailing paradigm or world picture, he would not intentionally try to go beyond the boundaries of what is considered true or plausible. And even if he is aware of the limitations of the scientific world picture and desires to transcend it, he does not have a clue how to do it."[4] In contrast, Kuhn pictures the much rarer and paradigm-transforming "revolutionary science," which changes the fundamental structures of science—either of a whole science (as achieved by Einstein, Newton, or Darwin) or, more often, a significant sub-specialty of a major science.[5] Science can be essentially transformed or re-directed by new theories, discoveries, or major technologies.

But how to balance a mindset that's open and unbiased with historic precedents? Here, there's a sharp difference in attitude. On the one hand, for example, Richard Feynman, Nobel laureate in physics, disdained reading journal articles, or anything regarding previous achievements, so that he could approach problems with a fresh, unbiased mind. Feynman was considered the leading intuitionist of his age. In contrast, there is Judah Folkman, the physician investigator at Harvard Medical School who introduced and developed the concept of angiogenesis (the development of new blood vessels in tumor growth and spread) over the past four decades. In his formative years, Folkman read all of the Nobel Prize lectures and acceptance speeches, hoping to uncover the patterns of creative thinking.

Over time, the idea has taken hold that advances can come about only through what's called "Big Science": highly funded, resource-rich investigations supported by modern technology in an ivory tower or an industrial complex.[6] Big Science is almost inevitably a type of normal science (since it tends to be predictable) and tends to be "applied" rather than pure or basic science. Further, since it depends on group collaboration, it values traits of collegiality and the ability to work congenially with others over creativity.[7]

Yet, while funding of research is important, what drives science is ideas. J. J. Thomson, the British Nobel laureate in physics who built the Cavendish Laboratory at Cambridge University into the greatest research institution in the world, used to say that if government patronage of science and technology had existed in the Stone Age, we'd all have wonderful stone tools today and no metals.[8]

In his farewell address on January 17, 1961, President Dwight Eisenhower famously cautioned the nation about the influence of the "military-industrial complex," coining a phrase that has become part of the political vernacular. However, in the same speech, he presciently warned that scientific and academic research might become too dependent on, and thus shaped by, government grants. He foresaw a situation in which "a government contract becomes virtually a substitute for intellectual curiosity." Many of the most essential medical discoveries came about through investigations that were driven by curiosity, creativity, and, often, a disregard for conventional wisdom. A concept may be abetted by technology, but the foremost stimulus is the idea, the insight.

Bruce Charlton of Newcastle University, UK, editor of the journal *Medical Hypotheses*, has used science Nobel Prizes over a recent sixty-year period as a metric for measuring revolutionary science. Although he allows that the small annual number of Nobel laureates means that many significant achievements go unrecognized, nonetheless, the perceived validity of these awards is high within the scientific community. He concludes that the United States, which sustains numerous elite research institutions, has become the only nation that supports revolutionary science on a large scale.[9]

THE ELUSIVE CRYSTAL BALL

In the medical literature of the past century or so, there are two outstanding guides offering timeless advice as to the nature of scientific investigation. The first, by the groundbreaking neurobiologist Santiago Ramón y Cajal, is still relevant since its original publication in 1897 and covers topics from which personality traits are valuable for an investigator to which social factors are conducive to scientific work.[10]

The second, published in 1979 by another Nobel laureate, Peter Medawar, is a thoughtful insight into the pitfalls and rewards of the scientific process.[11] The book deflates the myths surrounding scientists and makes observations about their psychological quirks and passion for inquiry. Medawar points out elsewhere: "It is impossible to predict new ideas—the ideas people are going to have in ten years' or ten minutes' time—and we're caught in a logical paradox the moment we try to do so. For to predict an idea is to have an idea, and if we have an idea it can no longer be the subject of a prediction."[12]

Nothing illustrates this more dramatically than the utterances of false prophets who, despite stunning breakthroughs in medicine, science, and technology over the past century and a half, have long trumpeted the end of advances.[13] Consider these quotes:

> *The abdomen, the chest, and the brain will be forever shut from the intrusions of the wise and humane surgeon.*
>
> —*Sir John Erichsen, British surgeon, later appointed Surgeon Extraordinary to Queen Victoria, 1873*

Virtually everything that needs knowing about the material universe has been learned, with only a few loose ends to be tidied up.

—Lord Kelvin, English physicist and president of
the Royal Society, 1896

Everything that can be invented has been invented.

—Charles H. Duell, commissioner of the U.S. Patent Office,
in a letter to President William McKinley urging
him to close the office, 1899

We can surely never hope to see the craft of surgery made much more perfect than it is today. We are at the end of a chapter.

—Berkeley George Moynihan, Leeds University Medical School, 1930

The energy produced by the breaking down of the atom is a very poor kind of thing. Anyone who looks for a source of power in the transformation of the atom is talking moonshine.

—Lord Ernest Rutherford, 1933

The great era of scientific discovery is over... Further research may yield no more great revelations or revolutions, but only incremental, diminishing returns.

—John Horgan, science journalist, 1997

Reality shows us that such statements border on farce. Clearly, scientific advances are accelerating at a furious pace and show no signs of slowing down.

A JURY OF ONE'S PEERS

These thoughts lead us to reconsider peer review by granting agencies and journals. We are accustomed to rely upon peers—teachers, referees, colleagues, competitors—in many activities. They point out deficiencies in our thinking and weaknesses in our project design. Medicine and science have not developed an alternative to peer review, but its flaws must be recognized.

Most science is an attempt to build upon prior knowledge while maintaining an outlook within the conventional wisdom. But as I noted in chapter 9, peer review tends to inhibit innovation. In a self-effacing editorial in 2003, *Nature* emphasized "the difficulties in assessing groundbreaking work" and acknowledged that the large number of "rejections experienced by Nobel winners...neatly illustrates the hurdles they had to overcome to publish their work." *Nature* lamely concluded by recommending "rejected authors who are convinced of the ground-breaking value of their controversial conclusions should persist."[14]

Peer review demands conformity of thinking and tends to reinforce dogmatism. "We can hardly expect a committee," said biologist and historian of science Garrett Hardin, "to acquiesce in the dethronement of tradition. Only an individual can do that."[15] Who on a review committee is a peer of a maverick, an individual who views a problem with fresh eyes? An applicant for a research grant is expected to have a clearly defined program over the next three to five years. Implicit is the assumption that nothing unforeseen will be discovered in the meantime and even if it were, it would not cause distraction from the approved line of research. How can a venture into the unknown offer predictability of results?

This view has not been lost on biochemists Stanley Cohen and Herbert Boyer, who won the Albany Medical Center Prize in Medicine and Biomedical Research, the nation's largest award in the field, in 2004 for their pioneering work to splice and recombine genes. After working at the University of California, San Francisco, Boyer founded the biotechnology company Genentech Inc. Cohen explained: "Herb and I didn't set out to invent genetic engineering" in the early 1970s. "Were we to make the same proposals today, some peer review committees would come to the conclusion that the research had too small a likelihood of success to merit support."[16]

Similarly, with scientific presentations—either at meetings or as journal articles—the true value of a contribution has been difficult to assess. Pioneering researchers have found that even when they are scheduled at major conferences to present their findings, their work has not been appropriately judged. The presentation may be relegated to an embarrassingly limited venue, even when the work leads to a major breakthrough.

Let us recall a few notorious examples:

- Pierre Deniker, a psychiatrist presenting the first positive results of clinical trials of Thorazine at a major congress in Paris in July 1952, was scheduled to speak at the end of the last session during the lunch hour to not more than twenty registrants. This new drug would shortly revolutionize psychiatry with an astounding impact: the release of hundreds of thousands of patients from mental hospitals.
- A few years later in September 1957, another psychiatrist, Roland Kuhn, presented the beneficial clinical results of a tricyclic anti-depressant at an international congress in Zurich to an audience of barely a dozen people. This new class of drugs would have a profound effect on millions of patients suffering from depression.
- Another clinical researcher, Robert Noble, found himself presenting his groundbreaking findings on successful cancer chemotherapy derived from a plant at midnight before a small cluster of registrants at a meeting of the New York Academy of Sciences in 1958.[17]

In each instance, these scientists' discoveries transformed the world.

Is it surprising that the core papers of a research front sometimes have difficulties in getting published because their original referees rejected them?[18] Let me briefly highlight a few examples:

- Rosalyn Yalow, in the Nuclear Medicine Department at the Bronx VA Hospital in New York, received a letter of rejection from a prestigious American journal. The journal did not want to publish her path-breaking submission, which led to a technique of measuring even minute amounts of substancers in the body. She was so bitter over this that twenty-two years later she cited the letter and even reproduced it in her 1977 Nobel Prize lecture. Yalow—a feminist to her core before feminism reached society's consciousness—suffered another indignity when her Nobel Prize was extensively covered in women's magazines. One magazine headlined an article, "She Cooks, She Cleans, She

Wins the Nobel Prize." (Obviously, little consciousness-raising had occurred since 1964, when Dorothy Hodgkin of Oxford University won the Nobel Prize for her determination by X-ray techniques of the structures of biologically important molecules and *The Daily Mail* headlined, "Grandmother Wins Nobel Prize.")

- A common complication of cirrhosis (fibrosis) of the liver—from alcoholism as well as other causes—is the development of engorgement of veins that may seriously bleed at multiple sites in the body. Joseph Rösch, a radiologist in Portland, Oregon, introduced a technique to alleviate the back-pressure in veins known as portal hypertension. His article was initially rejected by three surgical journals. I asked Rösch why. He answered, "They thought I was crazy!"[19]

- Barry Marshall, a young physician in Perth, Australia ("the wrong end of the wrong place"), far from the research centers in the Western Hemisphere, persisted for ten years against long-held dogma regarding the cause of peptic ulcer disease before his revolutionary concepts were widely accepted. Rebuffed by gastroenterologists, he was forced to seek an infectious disease conference in Brussels in September 1983 to present his startling observations on the bacterial association with ulcers. A report he first submitted to the *New England Journal of Medicine* was rejected and was then published in the *Lancet*.[20]

- Judah Folkman persevered for over four decades delineating the role of new blood supply recruited in tumor growth and spread, finding difficulty for many years in having his work published. The son of a rabbi, he had been given "Moses" as a middle name, but he never felt the irony in his forty-year-odyssey.[21]

Three researchers shared the 2007 Nobel Prize in medicine: Mario R. Capecchi of the University of Utah in Salt Lake City, Oliver Smithies of the University of North Carolina in Chapel Hill, and Sir Martin J. Evans of Cardiff University in Wales. They developed the immensely powerful "knockout" technology that allows scientists to create animal models of human disease in mice. The technique, also known as "gene targeting in mouse embryo-derived stem cells," is a superb new tool for finding out what any given gene does. By genetically engineering a strain of mice with the gene missing (or knocked out), researchers then watch to see what the mice can no longer do. The work has led to the breeding of "knock-out mice" and "knock-in mice"—animals with a single gene removed or inserted. While extending basic understanding, the breakthrough will lead to the development of new therapies.

The prize was particularly rewarding for Mario Capecchi, who was forced to live as a street urchin in northern Italy during World War II. He had to prove his scientific peers wrong after they rejected his initial grant to the NIH in 1980, saying his project was not feasible. What he proposed was manipulating mouse genes to help model disease. When he applied for the grant four years later, he was told, "We are glad you didn't follow our advice." Sir Martin Evans also said his scientific career was an upward struggle. In applying for grants, he was told many of his ideas were premature and could not be done. "Then five years later," he commented, "I find everyone is doing the same thing."[22]

All these experiences raise a fundamental question. If there is such limitation in recognizing groundbreaking initiatives, how can the process of establishing credit and priority—recognition and reward—be flawless? The fact is, it's not.

BAD DECISIONS

As assiduous as the Nobel committee's deliberations are, more than a few of their judgments have been off the mark. The most embarrassing occurred in the case of the Danish pathologist Johannes Fibiger, awarded the Nobel Prize in medicine in 1926 for a fallacious conclusion. This reflects the trap for scientists lurking in the common logical fallacy *post hoc, ergo propter hoc*—"After it, therefore because of it." This is the tendency to attribute an occurrence to whatever preceded it in a chronological arrangement of events. Fibiger, for instance, discovered roundworm parasites in the stomach cancers of rats. He was convinced that the rats ate the larvae of the parasite in cockroaches, and this is what brought about the cancer. He presented experimental work in support of this theory. His results were subsequently never confirmed, and the theory was justifiably abandoned. Scientists later concluded that most of

Fibiger's laboratory rats died of tumors caused by a vitamin deficiency. It was forty years before a chagrined Nobel Assembly dared again to award the Prize for cancer research.

The most regrettable error in judgment is the Nobel Prize for medicine awarded to António Egas Moniz in 1949 for the procedure of blindly cutting nerve pathways in the brain in patients with anxiety states or "disturbing" social behavior. The technique was modified and practiced by Walter Freeman, a neurologist, in the United States; often referred to as "ice pick lobotomy," critics denounced it as "not an operation, but a mutilation."

But more important than mistaken recognition are several deserving cases that went unrewarded:

> *"I have never once doubted the universality of this law,*
> *because it could not possibly be the result of chance."*

In what is generally considered the most masterful stroke of pattern perception in the history of science, Dmitri Mendeleev, professor of chemistry at St. Petersburg University, created the periodic table of elements in 1869. It has been displayed since in every chemistry classroom. For years, he had brooded on how the chemical elements might be classified. When seized with a problem or an idea, Mendeleev characteristically attacked it with great concentration. In a period of almost ceaseless intensity in 1869, he painstakingly arranged and rearranged a mass of data, seeking an encompassing order between the elements. To a friend at one point, he lamented, "It's all formed in my head, but I can't express it in the table."

Mendeleev liked playing solitaire, arranging the cards in rows of suits in descending order. In a stroke of intuition, he pulled out sixty-three white cards and on each wrote the chemical symbol of a known element, its atomic weight, and a short list of its characteristic properties. Analogy to the game of solitaire, he thought, might uncover a pattern similar to sorting out cards according to suit—"similar elements"—and denomination—"similar atomic weight." Could there be a relationship between an element's chemical structure and its physical properties?

Initially, Mendeleev could discern no overall pattern. It was evident that groups with similar properties, such as the halogens, fluorine, chlorine, bromine, and iodine, had widely different atomic weights. Yet he felt that he was on the verge of unfolding a great secret. Overcome by exhaustion, he fell asleep at his desk. On the night of February 16, 1869, the insight regarding the overall guiding principle came to him in a dream. In his own words: "I saw in a dream a table where all the elements fell into place as required. Awakening, I immediately wrote it down on a piece of paper."[23] He realized that if the elements are listed in order of ascending atomic weights, certain similar properties are repeated at regular or periodic intervals (rather like musical octaves), particularly in the elements' ability to combine with other elements. He published his periodic table of the elements two weeks later in the *Journal of the Russian Chemical Society.*

Mendeleev's boldness and self-confidence furthered the breathtaking nature of this contribution. When certain accepted facts of some elements did not fit securely into his pattern, he argued that they were erroneous, and, in case after case, he was proved right. He reserved several blanks in his table for elements "as yet unknown." His predictions for their existence were affirmed within a few years. In 1875, a French chemist discovered gallium, named after the Latin word for France. This was followed by other scientists' discoveries of the elements scandium and germanium (similarly labeled with an eye to nationalistic pride). Explaining what can already be seen is certainly creative, but using that pattern to anticipate as yet undisclosed findings adds immeasurable value.

Half a century later, the rationale underlying Mendeleev's contribution would gain a foundation in the concept of electrons orbiting an atom's nucleus. Mendeleev had the courage of a prophet: "I have never once doubted the universality of this law, because it could not possibly be the result of chance."[24] Oliver Sacks has recorded his awe as a youth on encountering Mendeleev's work: "When I first saw the Periodic Table, it hit me with the force of revelation—it embodied, I was convinced, eternal truths...I thought of Mendeleev as a sort of Moses, bearing the tablets of the God-given Periodic Law."[25]

Mendeleev was nominated in both 1905 and 1906 for the Nobel Prize in chemistry, but lost out because one committee member, an influential member of the Academy, insisted that his periodic system was thirty-five years too old, despite the fact that a newly established group of elements proved the validity of the periodic law. Only later was the strong-willed Academy member's stance found to be based on a personal grudge.[26] As Burton Feldman, a historian of the Nobel Prizes, pointed out, the awards have been presented at least a few times for work that dated back several decades.[27] Ironically, the committee subsequently honored several people for the discovery of new elements, precisely what Mendeleev had predicted. Unfortunately, he didn't live long enough for his due recognition; he died in 1907.

In a fitting tribute, a new transuranium element, Number 101, created in 1955, was named *mendelevium*.

"My name ran right along with Professor Millikan's in the newspaper."

One graduate student's claim for final recognition of his contribution to research for which his mentor had received the Nobel Prize was heard, after many years, as a voice from the grave. Harvey Fletcher, a twenty-five-year-old graduate student, worked with his thesis adviser, Robert Millikan, at the University of Chicago. Years later, Fletcher instructed a friend to publish his personal memoir only posthumously, so it would be clear that Fletcher had nothing to gain from its publication.

When Fletcher joined Millikan's lab in 1909, the existence of the electron was becoming widely accepted by experimentalists as more than a heuristic device. Values were sought for the magnitude of the electron's charge. Millikan was working on a simple experiment that imposed an electric field upon a cloud of tiny water drops in a chamber. The droplets would soon fall under the influence of gravity. By modifying the intensity of the field, Millikan attempted to suspend a selected droplet. The electrical charge in the droplet could be calculated from the fall speed and the intensity of the field required to stop its fall. But because the water evaporated quickly, only a rough estimate of the charge on different "balanced" water drops could be made. This was essentially repeating the experiment done by a physicist in England.

Harvey Fletcher alleges that it was he who suggested using oil droplets, and he designed and constructed the necessary apparatus for the experiment, ingenious in its simplicity. Millikan urged Fletcher to devote his Ph.D. thesis to this project.

Gerald Holton, a historian of science, calls this a "process of significant maturation" because sets of data could now be obtained on the risings and fallings of a single oil drop. Holton reviewed Millikan's laboratory notebooks and notes that he expressed his exaltation at the data in the corner of one page: "Beauty. *Publish* this surely, *beautiful!*"[28] After much work with many variations, Millikan and Fletcher finally obtained the now widely accepted value of *e*. The precise measurement of the electron, central to the workings of the atom, became the benchmark against which other particles were gauged.

Millikan and Fletcher published five papers over the next nine months. But the most important was the first, in 1910, announcing the precise measurement of the electron.[29] Fletcher states that he wrote the majority of this pivotal paper and expected to be a joint author, but Millikan demanded exclusive authorship. The paper, which attests to Fletcher as a co-experimenter, led to Millikan receiving the Nobel Prize for physics in 1923.

Fletcher graduated with a Ph.D. in physics summa cum laude in 1911 and went on to direct acoustical and, later, physical research at Bell Laboratories from 1925 to 1952. Millikan became director of the California Institute of Technology.

Fletcher recalls, "The papers were full of this wonderful discovery. It was the first real publicity that I had ever received. My name ran right along with Professor Millikan's in the newspaper." At the end of his article, published posthumously in 1982,[30] Fletcher acknowledges Millikan's friendship and generosity, so the reader may feel that it is a poignant recollection. But, in the last paragraph, Fletcher indicates that many felt that Millikan had treated him unfairly. Fletcher's disclaimer that there was no personal interest motivating the article's publication is a bit disingenuous since it is clearly written with an eye to history.

"Perhaps you can come up with some sort of fantastic conclusion."

Among deserving scientists ignored by the Nobel committee was Lise Meitner, whom Albert Einstein fondly referred to as "our Marie Curie." A shy, diminutive woman, Meitner closely collaborated with the chemist Otto Hahn for over thirty years at the prestigious Kaiser Wilhelm Institute (KWI) for Chemistry in Berlin-Dahlem. Despite the severe gender discriminations of the time, especially in Germany, Meitner—having achieved a doctoral degree in physics at the University of Vienna—was given her own physics section at the KWI in 1917.

In 1934, she convinced Hahn to join with her to investigate the very heart of the atom, its nucleus, and seek elements beyond uranium, then the heaviest atom known. At the time, this was considered basic research for the probable honor of a Nobel Prize, and no one suspected that it would culminate in nuclear weapons. Meitner was the intellectual leader of the team, which found itself in competition with Irene Joliot-Curie and her collaborators in Paris. Hahn and a gifted analytic chemist, Fritz Strassmann, performed the painstaking chemical analyses to separate and analyze the minute radioactive specimens. By bombarding uranium with neutron particles, they encountered radioactive byproducts that could not be easily identified.

After Austria was annexed by Germany in 1938, Meitner, an Austrian Jew, was forced to flee Germany for Sweden. Before she left, Otto Hahn, who refused to join the Nazi party and had become a valued friend, gave her a diamond ring he had inherited from his mother that was to be used to bribe the frontier guards if required. Meitner was fifty-nine and took up a post in Stockholm in the laboratory of Manne Siegbahn, a Nobel laureate in physics. Within a few months, on November 13 and 14—just after Kristallnacht marked an escalation in Nazi measures against Jews and dissidents—Hahn and Meitner met clandestinely in Copenhagen to plan a new round of experiments, and they subsequently exchanged a series of letters. Meitner knew that the results Hahn was sending her could not be correct and urged refinement of the most critical experiments and new measurements. At her direction, Hahn and Strassmann came to realize to their amazement that the products of the neutron-bombardment weren't heavier than uranium but lighter. Baffled by this remarkable conclusion, Hahn wrote to Meitner on December 19: "Perhaps you can come up with some sort of fantastic conclusion."[31] Within days, collaborating with her nephew Otto Robert Frisch, also a noted physicist living in Sweden because of Jewish persecution, she worked out a theoretical model of nuclear fission—of how the nucleus of an atom could be split into smaller parts. Meitner also first realized that Einstein's famous equation, $E = MC^2$, explained the tremendous releases of energy in atomic decay by the conversion of the mass into energy.

Hahn, along with Strassman, published the chemical evidence for fission in January 1939, without suggesting any mechanism for splitting the atom nor considering the release of energy.[32] Neither did they list Meitner as co-author, a move she understood at the time given the political climate. When they submitted the paper, they communicated their results to her in a letter. Frisch discussed the conclusions that Meitner and he had come to with Niels Bohr, who immediately accepted their results and allegedly exclaimed: "Oh, what fools we have been! We ought to have seen that before." Meitner and Frisch published a correct interpretation of the results in a landmark letter to the journal *Nature,* introducing the term nuclear fission, by analogy with the process by which a living cell splits in two.[33]

Bohr indiscreetly leaked the information before the paper was in print, generating a frenzy of intellectual and experimental activity by other physicists. Meitner and Frisch's original insight, arrived at with such difficulty, suffered the consequence of being viewed as almost intuitively obvious—once understood.

Leading American physicists, recognizing that the knowledge of such huge explosive potential was in German hands, persuaded Albert Einstein to write a warning letter to President Franklin D. Roosevelt. This led directly to the establishment of the Manhattan Project.

Meitner's biographer underscores that Hahn had been distancing himself from his longtime collaboration even before Meitner escaped. But in his defense, he was under strong pressure from the Nazis to minimize Meitner's role. It is his conduct after the war that is less than noble. He maintained that his chemical experiments had been neither inspired nor guided by Meitner, and he took sole credit, claiming that the discovery was his alone. Whereas Meitner and Hahn had always shared the credit on their joint efforts, her name was missing from the key experimental paper on nuclear fission, and she came to be mistakenly viewed as his junior assistant. In 1944, Hahn alone received the Nobel Prize in chemistry. Robert Marc Friedman traces the complex dissensions among the Nobel chemistry and physics committees, the influence of Meitner's mentor Siegbahn, and Sweden's close relations with Germany before and during the war years and later attitudes toward American science. He concludes: "The mid-1940s Nobel prizes were not awarded on the basis of recognizing merit; instead, they had become to a great extent instruments in the politics of science."[34] Lise Meitner was denied her due recognition.

"... it is possible to induce predictable and hereditary changes in cells."

The discovery in 1944 that DNA could cause genetic transformations among strains of certain bacteria "marked the opening of the contemporary era of genetics," writes Joshua Lederberg, "its molecular phase."[35] Oswald Avery, a researcher at the Rockefeller Institute in New York City, demonstrated that the hereditary material—the stuff of genes—is made of nucleic acid and not, as long thought, protein. Nucleic acids were known since the latter part of the nineteenth century, and DNA (deoxyribonucleic acid) and RNA (ribonucleic acid) were distinguished in the 1920s. Both contain, among other components, the sugar ribose plus carbohydrates; "deoxy" means the DNA lacks one oxygen atom in its sugar structure. Although suspected of playing a role in metabolism, nucleic acids seemed simple in composition, and their specific biological role had remained unclear. Avery's evidence came from experiments developing antisera against pneumococcal bacteria.

In a famous letter to his brother Roy, Avery wrote that his results might indeed mean that nucleic acid was the substance by which "it is possible to induce *predictable* and *hereditary* changes in cells." He added, "It's lots of fun to blow bubbles—but it's wise to prick them yourself before someone else tries to."[36] The findings fell mostly on deaf ears. For publication, Avery's characteristic caution prevailed. In 1944, together with two colleagues, he published a landmark paper in the *Journal of Experimental Medicine*. He was sixty-six years old. Its unprepossessing title was "Studies on the Chemical Nature of the Substance Inducing Transformation of Pneumococcal Types." In this self-effacing and hyper-cautious way, the proof that nucleic acid, not protein, composed the gene slid into the scientific literature. Avery restrained himself from going so far as to suggest that DNA would constitute the chemical basis of heredity. One writer observed that he was "almost neurotically reluctant to claim that DNA was genes and genes were simply DNA."[37]

Avery stood a good chance of receiving a Nobel Prize in the 1950s. Findings by other researchers added credence to his work. Recognition that the structure of DNA was "double-helical" and ladder-like—suggesting "a possible copying mechanism for the genetic material"—was finally delivered by Francis Crick and James Watson, in a letter published in *Nature* in 1953.

Many agree that the Nobel committee's failure to award Avery the prize is one of its most glaring omissions. It decided to wait until more became known about the transformation of DNA, and one member stubbornly refused to admit DNA was anything more than structural support for the genetic material contained in proteins. Within a few years, the verdict was in: "Avery's discovery... represents one of the most important achievements in genetics and it is to be regretted that he did not receive the Nobel Prize."[38] Oswald Avery, a quiet and reserved man, died in 1955.

"Jocelyn was a jolly good girl, but she was just doing her job."

In 1965, Jocelyn Bell began work on her Ph.D. at Cambridge University under the direction of Anthony Hewish. She was an energetic twenty-two-year-old Irish Quaker with a B.S. in physics. Quasars—a distant energy source that gives off vast amounts of radiation, including radio waves and X-rays—had recently been discovered, and Hewish designed a large radio telescope to identify and track them. The phrase "radio telescope" conjures the wrong image in most people's minds. What Hewish had in mind was a construction over four and a half acres of about 1,000 posts stuck into the ground in neat rows with over a hundred miles of wire and cable stretched across them to create about 2,000 dipoles. The telescope scanned across the sky with time due to the rotation of the Earth.

Bell's first two years at Cambridge were spent, with the help of others, constructing the 81.5 megahertz instrument in a field a few miles outside of Cambridge, as part of the Mullard Radio Astronomy Observatory. The slight young woman became adept at wielding a twenty-pound sledge hammer. The telescope went into operation in July 1967, and it was Jocelyn Bell's responsibility to operate it and analyze the data, with supervision from Hewish. The output appeared on about 400 feet of three-track chart paper over the four days that it took for each complete coverage of the sky. Bell's job required intense concentration to scan the chart by eye, discarding the signals that came from man-made sources of interference, and mapping the signals that were true twinkling radio sources. As Nicholas Wade has recounted the events,[39] by October, Bell was 1,000 feet of chart behind. Nevertheless, she continued to scrupulously examine the tracings inch by inch.

Quasars scintillate, that is, fluctuate in the intensity of their radio emissions. However, in October, Bell noticed what she described as "a bit of 'scruff' on the records," a signal that occupied about one inch of the 400 feet of chart. It "did not look exactly like a scintillating source and yet did not look exactly like man-made interference either." Remarkably, she realized that she had seen this before on the same part of the records from the same patch of sky. "It seemed to be keeping pace...with the rotation of the stars." That single instant of recognition would herald a major discovery. Although no one had yet realized it, Jocelyn Bell had detected the first evidence of a pulsar. When she recorded another similar observation, she noted that the series of pulses were evenly spaced and were $1^1/_3$ seconds apart. The short repetition time was totally unexpected.

Hewish's first reaction was that these signals must be man-made. But observations by other radio astronomers established that the source was well outside the solar system. As the puzzle deepened, Bell and Hewish considered the possibility that this regular signal coming from so far away could actually be some form of life in the far-distant universe. Thus arose briefly the fanciful thesis that Little Green Men in a remote galaxy were transmitting periodic signals in a sort of intergalactic social network.

This could be discarded with the finding just before Christmas of another "bit of scruff" with a period of about one and a quarter seconds originating from a completely different part of the sky. "It was very unlikely," Bell concluded, "that two lots of little green men would both choose the same, improbable frequency, and the same time, to try signaling to the same planet Earth." The source must be an object spinning at great speed and sending out radio waves in a beam, like light from a lighthouse. Bell had, in fact, discovered pulsars.

Hewish and Bell published the discovery of pulsars in the February 1968 issue of *Nature*. Bell's name was second on the list of five authors.[40] This would clearly indicate to the article's readership that the discoverer of pulsars was Hewish, along with four members of his group. Further studies by groups of astronomers around the world confirmed that the signals were coming from rapidly rotating neutron stars, supernova remnants. A supernova appears as a gigantic explosion. Most of the star's mass falls inward, compressing itself so tightly that the individual atoms collapse, with the electrons and the protons coalescing to form neutrons. The term pulsar is an abbreviation for pulsating radio star or rapidly pulsating radio sources. Pulsars, showing that one of the end states of stellar evolution could be neutron stars, are widely recognized as one of the great astronomical discoveries of the century.

In 1973, the Albert Michelson Medal of the Franklin Institute of Philadelphia was awarded jointly to Anthony Hewish and Jocelyn Bell.

The following year, 1974, Hewish was awarded the Nobel Prize for physics for the discovery of pulsars; Jocelyn Bell was not included as a co-recipient. This was openly and bluntly condemned in *The Times* of London as a "scandal" by Hewish's fellow astronomer, Sir Fred Hoyle of the Institute of Astronomy at Cambridge. He asserted: "[Her] achievement came from a willingness to contemplate as a serious possibility a phenomenon that all past experience suggested was impossible." In response, Hewish contended that Bell had been using his telescope to make a sky survey that he had initiated. "Jocelyn was a jolly good girl but she was just doing her job," the Nobel laureate explained. "She noticed this source was doing this thing. If she hadn't noticed it, it would have been negligent."

Yet, Bell's perspicacity and self-motivated pursuit of the finding are highlighted by this key passage of her account:

> The discovery was almost totally unexpected. We learned later of a radio astronomer at another observatory—I won't say who or where—who several years earlier was observing a portion of the sky to the right of Orion, northward, where we now know there to be a pulsar. And he saw his pen [on the recording device] begin to jiggle. And he was about to go home for the day, and thought his equipment was misbehaving. And he kicked the table and the pen stopped jiggling.[41]

And what was Jocelyn Bell's reaction to having been overlooked? Surprisingly humble and sublimely accepting. She countered Hoyle's argument by serenely stating that "[graduate] students don't win Nobel Prizes."

She expanded her views upon this in 1977:

> It has been suggested that I should have had a part in the Nobel Prize
> awarded to Tony Hewish for the discovery of pulsars. There are several
> comments that I would like to make on this: First, demarcation disputes
> between supervisors and student are always difficult, probably impossible
> to resolve. Secondly, it is the supervisor who has the final responsibility for
> the success or failure of the project...Thirdly, I believe it would demean
> Nobel Prizes if they were awarded to research students, except in very
> exceptional cases, and I do not believe this is one of them. Finally, I am not
> myself upset about it—after all, I am in good company, am I not?[42]

Jocelyn Bell received her Ph.D. in radio astronomy from Cambridge
University in 1968 (her thesis was not based on pulsars). She married the
same year and changed her name to Burnell. She went on to a distinguished
career, receiving many honors and awards.

THE MENTOR-STUDENT
RELATIONSHIP REVISITED

Bell's case raises a question of ethics: when a laboratory director sets a graduate student to a specific task and the student astutely notices an unexpected
finding, an anomaly, and pursues it, along with the director's involvement,
to a remarkable conclusion, who deserves the credit?

It is fairly easy to condemn Hewish, but the mentor-student relationship
is one with certain benefits for graduate students. By cooperating in a joint
study, the student is assured of income, given a task that is to be accepted as
professionally rewarding, and provided a thesis project, data, a place to work,
secretarial assistance, etc. In return, the student can, in fact, expect that the
mentor will continue to look out for him/her. When it comes to finding a
position, a patron is a good person to know—for letters of recommendation,
the networks through which so many jobs are landed. And then there is
the possibility of a continuing long-term relationship, with favorable recommendations on grant applications.

Even today, decades after the Waksman-Schatz conflict flared, little has changed in the relationship between the laboratory chief and the staff of graduate students and postdocs. Most of the approximately 150,000 Ph.D. life scientists in the United States work at colleges and universities, where the federal government now is the major single source of biomedical research funding. A typical research laboratory is run by a principal investigator (PI) who is responsible for choosing research topics, raising money, juggling budgets, and managing postdocs and graduate students. The latter are motivated by the hope of achieving an independent research career and making important contributions, rather than monetary incentives.[43]

As seen through the graduate students' eyes, their plight has changed little over the years. Excerpts from "The Grad Student Rap" by Adam J. Ruben, who recently completed seven years working on a Ph.D. in molecular biology at Johns Hopkins University, take a painfully humorous shot at the process:

> Takin' student loans just to have some money.
> Payin' them off 'til my kids are 20.
> Every passing day makes me afraid.
> Hey, just think of this as 23rd grade
> (chorus)
> *I'm a grad student*
> *In a community of scholars.*
> *I'm a grad student*
> *I make dozens of dollars!*
> *I'm a grad student*
> *And I'm sure you'll agree*
> *That's why they're givin' me the third degree.*[44]

Students and postdocs depend on the laboratory for education, career development, and income. Increasingly, the tournament nature of research in the biosciences fosters intense competition: the chance of winning a big prize—an independent research career, tenure, a named chair, scientific renown, awards.

Distinguishing between full-fledged scientific collaboration and supervised research assistance, between irreplaceable and replaceable scientific contributions to prize-winning research, certainly may be difficult. The PI serves as the team's spokesperson in presenting the research findings to scientific meetings and the media. Whether by marginalizing or acknowledging to some degree the contributions of others, primary credit flows to that individual. This became the source of friction in the instances of Frederick Banting with John Macleod, René Dubos with Oswald Avery, Harry Loomer and John Saunders with Nathan Kline, and Georges Köhler with César Milstein.

In stark contrast to Jocelyn Bell's restraint in the face of great temptation to claim a share in the Nobel for herself, there is the case of Candace Pert, a graduate student who worked in the laboratory of Solomon Snyder at Johns Hopkins University. In 1978, Snyder was one of three investigators who won the Lasker Award for Basic Medical Research for their discoveries of opiate receptors in the brain. At the time of the award, Pert was a research scientist at the National Institute of Mental Health (NIMH). She loudly complained that she was unfairly neglected. Many of her colleagues at the NIMH supported her. In a letter to the Lasker committee, she wrote, "I was angry and upset to be excluded from this year's Award...As Dr. Snyder's graduate student, I played a key role in initiating this research and following it up." Pert strongly believed that she originated the research leading to the isolation of opiate receivers in the brain. Snyder maintained that he began the project.

In any event, Pert is the first author on the original papers dealing with identification of the opiate receptors and also on many of the other achievements for which Snyder was cited. Snyder allowed that "it would have been appropriate if Pert had shared the award with him" and asked members of the jury of scientists who selected the award winners to consider including Pert after all, a request that was refused.[45]

FOLLOWING THE MONEY

Toward the end of World War II, choosing a scientific career was far from any motivation of getting rich. American academic scientists started out on about $2,000 a year—the rough equivalent of $17,000 these days—while a few professors at the peak of their careers commanded as much as $10,000.

The American scientist, a writer in *Science* observed in 1953, "is primarily interested in what he can do for science, not in what science can do for him." Around the same time, Karl Compton, a physicist and president of MIT, said of scientists "I don't know of any other group that has less interest in monetary gain." In letters to Waksman, Schatz repeatedly emphasized that he had no regard for money. Only when he felt exploited and betrayed did he pursue what he then believed to be his entitlement.

Traditional scientific values could not remain unaffected, however, as they became challenged with the enormous wealth directed to American science from early in the Cold War in the 1960s. An outstanding example is represented by the biotech industry, which was created in 1976 when Herbert Boyer, a biochemist at the UCSF, working with the Stanford geneticist Stanley Cohen, developed some elegant recombinant DNA technologies that immediately suggested enormous value to the pharmaceutical industry. As was normal at the time, intellectual property rights were assigned to the universities, from the licensing of which they eventually derived about $200 million before the patents expired in 1997. Some universities allow faculty members to hold patents; some give them joint interests in private sector ventures. With investment from venture capital, Boyer and Cohen incorporated a company known as Genentech, whose current market capitalization of $85.1 billion makes it the biggest biotech company in the world. This spinning-out of science from universities to entrepreneurial companies established new practical and moral possibilities for biomedical scientists: new ways of making large sums of money and new institutional forms for doing science.

SHARED TRAITS

Pondering the righteous indignation of researchers who feel strongly that their contribution has been trivialized, it is easy to draw up a list of qualities that they share.

Although decades apart, the similarities in actions and responses between Albert Schatz and Raymond Damadian are instructive. Once aroused, both were intrepid regardless of consequences. Both resorted to ruses and stratagems to advance their positions. Both sought support from previous Nobel laureates and other prominent scientists. Both were viewed as usurpers. Schatz petitioned the Swedish king, and Damadian argued his case before the public. And tragically both suffered the consequences of whistle-blowing: Schatz was blacklisted, never again to find a position in an academic laboratory, and Damadian was excommunicated by the NMR community, only to then be looked down upon for his role as corporate executive. The most telling shared traits are actions taken at the risk of their academic careers.

In contrast, comparing Selman Waksman and Paul Lauterbur, each was recognized as an authority at the nexus of a new burgeoning field. In both cases, the Nobel committee carefully chose wording to avoid conflict, emphasizing in Waksman's case not the discovery but the *methodology* of deriving streptomycin and in Lauterbur's case the *imaging* breakthrough derived from NMR.

It is evident that to those who feel themselves victims of injustice, the issue of maintaining their identification to the discovery becomes paramount and characteristically obsessive. They are compelled to act, with little regard for possible consequences. Propelled by a powerful feeling of righteousness, heedless of the harsh and inevitable reprisals, they take on the role of a whistle-blower.

Whistle-blowing generally deals with corruption in government and corporations.[46] Movies have depicted the struggles of a lone individual bravely confronting the unethical practices of powerful forces. These include *Serpico* (1973) regarding police corruption in New York City, *Silkwood* (1983) dealing with safety and health hazards in a nuclear plant, and *The Insider* (1999) about the tobacco companies' manipulation of the addictive properties of nicotine. Rarely considered are those who blow the whistle on misconduct directed solely against themselves.[47] In science, such incidents of unethical behavior underscore the essential role that proper attribution plays. Recognition provides both a major motivation and reward. Heidi Weissmann, Albert Schatz, and Raymond Damadian could each embrace the phrase, "This is *my* work, *my* legacy, *my* place in history, *my* bid for immortality."

Yet the outrage is so strong that little anticipation is made of the likely reactions, the falsity of the trust that someone in charge will do the right thing, and the need to fight a long battle for vindication. As events proved, they would be labeled as malcontents. An informal blacklist, occurring typically in tight-knit fields, aborted the careers of Schatz and Weissmann. Damadian was ostracized by the NMR community of chemists. It is a key element of whistle-blowing that rarely do individuals get demoted or fired by an organization for reporting the misbehavior of subordinates, but they invariably face vindictive retaliation when they challenge either their local institution or the external "community." This act of defiance of norms is often professional suicide and demonstrates to others the high cost of nonconformity.

Weissmann bristled over her authorship being "whited out." In her case, the dean disbanded a sympathetic faculty senate committee, a member of the university's board known for his empathy for the wronged and persecuted never responded to a personal appeal,[48] and security guards escorted her from her locked office off the medical center's premises. This latter act may be considered not just a degradation ceremony but a form of public execution.[49]

Schatz, with the high personal ideal "to be true to myself," struggled for fifty years to assert his claim. Damadian, fearing for over thirty years that he would be "written out of history," ultimately sought the loudest whistle, appealing—if not demanding—support and review by the Nobel Prize committee.

C. Fred Alford, the author of a definitive book on the subject, encapsulates the central dilemma: "Whistle-blowing is about trading off one dread for another. To remain within the system is to risk the dread of becoming dead to oneself. To step outside the system is to risk the dread of becoming dead to others."[50]

Vindication occurs only when the unethical practices are fully exposed and the offending individual or organization is forced to take corrective action. Still, one can question whether Albert Schatz receiving the Rutgers Medal, Heidi Weissmann receiving a court-directed financial compensation, or Margot O'Toole receiving a public apology from David Baltimore were fully gratified by the sense that their reputation and integrity had been restored. Nonetheless, whistle-blowers experience a substantial transformation of their lives, with loss of faith in society's major institutions, feelings of alienation and powerlessness, and bitter memories.

IN POPULAR CULTURE

The nuances of the issues determining scientific credit have been explored recently in plays and movies. The 2007 Broadway play *The Farnsworth Invention* is a story about the creation of television that pitted guileless inventor Philo T. Farnsworth against NBC founder David Sarnoff, a cold-blooded visionary. In the 2008 movie *Flash of Genius*, Robert Kearns, an electrical engineer and college professor in Detroit, invents and patents the intermittent windshield wiper in the late 1960s. The Ford Motor Company adopts his system without paying him or giving him credit. He sues Ford and then Chrysler for patent infringement and ultimately wins a settlement totaling $18 million.

As I mentioned in chapter 3, Rosalind Franklin's successive X-ray images provided the key to the helical structure of DNA in the breathless competitive race by James Watson and Francis Crick. In the 2009 off-Broadway play *Photograph 51*, by Anna Ziegler, Franklin is described as "two steps away from the solution. She just didn't know it."

In the 2010 movie *The Social Network*, Facebook's Mark Zuckerberg disdainfully confronts the two fellow Harvard students who claim he stole their idea with the slam-dunk statement, "If you guys were the inventors of Facebook, you'd have invented Facebook."

EPILOGUE

That science is not everything should not blind us to the fact that it is the very best of what we do have.

—*Henry H. Bauer*

Harold Varmus once described how strongly research gripped him: "It's an addiction. It's a drug. It's a craving. I have to have it."[1] Scientific research is certainly a noble endeavor, but not a holy one. It would be too facile to shrug off conflicts over credit as inevitable consequences of human nature, predicated on folly, greed, and hubris. Recognizing human nature as a constant, we must identify the sources of such crises and seek resolutions in changes. The starting point for our understanding is that the need for recognition and reward—to be esteemed by fellow scientists—is as strong a motivation as the quest to conquer the unknown.

In closing, I offer some suggestions on how conflicts such as the ones I have elucidated throughout this book can be avoided or at least mitigated. Reforms need to be made in:

Awareness of These Issues

Many scientists are in denial that these problems exist unless they are personally touched by them. Researchers may be so highly focused in their particular discipline that they remain ignorant of crises over priority and credit (and fraud as well) in other fields. Within the scientific community, such incidents may be seen as personal disputes or individual aberrations, acts that can have no significance for science as a whole. But to ignore that each incident strains science's public credibility is foolhardy.

Science has traditionally been founded on the bedrock of the "scientific method": systematic observation or experiment leading to hypotheses. In time, with validation based on further work, an overarching theory is formulated. In truth, despite appearances to the contrary, scientists in practice do not generally follow the scientific method.[2] Certainly, astronomers employ approaches different from biologists, chemists from geologists. The theory may be cast initially, awaiting only a critical observation to prove its infallibility. Einstein's theory of general relativity postulated few predictions that could then be tested by experiments. One of its central concepts was that space is curved and that this would cause light to bend around massive bodies even more than the effect of strong gravitational fields. This needed to be confirmed during a total eclipse of the sun. Several attempts were made by various national teams over the years, but without success. Finally, in November 1919, this concept was dramatically confirmed by two British teams, who measured how much the viewable light from stars bent as it passed near the sun during the eclipse. The evolutionary biologist Stephen Jay Gould has aptly framed the picture:

> Facts and theories are not rungs in a hierarchy of increasing certainty. Facts are the world's data. Theories are structures of ideas that explain and interpret facts. Facts do not go away while scientists debate theories to explain them. Einstein's theory of gravitation replaced Newton's but apples did not suspend themselves in midair pending the outcome.

Recently, the element of serendipity has been recognized as a pervasive dynamic influence in many major scientific breakthroughs. Indeed, serendipitous circumstances often spark scientific revolutions, including the discovery of the X-ray in 1895, the development of chemotherapy and psychotropic (mood-altering) drugs, and major cardiovascular advances like imaging of the heart's coronary arteries, enabling bypass grafts.[3]

Reflecting on his award-winning work in the co-discovery of stem cells, Ernest McCulloch was emphatic that he had always thought that the idea of "the scientific process" was overblown and that "typically a successful scientist may start with an experimental design but then makes an unexpected observation that leads a prepared mind to follow a chance event."[4] The physician-administrator-essayist Lewis Thomas made a similar assessment of the scientific method:

> I have never been quite clear in my mind about what this means. "Method" has the sound of an orderly, preordained, step-by-step process...I do not believe it really works that way most of the time...More often than not, the step-by-step process begins to come apart, because of what almost always seemed a piece of luck, good or bad, for the scientist; something unpredicted and surprising turned up, forcing the work to veer off in a different direction. Surprise is what scientists live for...The very best ones revel in surprise, dance in the presence of astonishment.[5]

If science is to be self-governing, it must govern itself consistently, fairly, and openly. The unintended consequences of the David Baltimore affair brought the wrath of the government and press down on the conduct. But as incident after incident accumulates, it becomes clear that an informed public will continue to allow the scientific community to govern itself only if it disseminates a publicly stated code of professional conduct, ethical guidelines to which the profession holds its members.[6]

Consistency in Attribution of Authorship

Authorship has been likened to a coin, with the two sides being credit and accountability.[7] Yet, the attribution of scientific credit based on authorship does not always fit the reality of research practices. Modern scientists face specific dilemmas. Seldom today do we see a single scientist independently conceive, plan, perform, and publish a body of research. In a group effort, the identification of the lead scientist for a particular segment of research can be ambiguous. Understandably, this may lead to difficulties in allocation of credit. Indeed, the most frequently discussed authorship issues are: Who should be a co-author of the manuscript? Who should be first author? What should be the order of authorship? Not surprisingly, such problems are widespread.[8]

Definitions of scientific authorship are not codified in a corpus of doctrine like intellectual property law.[9] However, the Committee on Publication Ethics in the United Kingdom, the International Committee of Medical Journal Editors (ICMJE, an influential body representing hundreds of biomedical journals), the NIH, and several universities and societies have reflected on and established guidelines pertaining to scientific authorship issues. But this doesn't necessarily solve the problem since there are variations among the guidelines.[10] For example, the senior laboratory scientist who may have been responsible for the initial overall concept under which the research was done as well as for writing the grant proposal that ultimately funded the research might be included under NIH guidelines but not under ICMJE guidelines.

The order of which author is listed first and which last as on research articles is also unclear. According to some, authorship order should be a joint decision of the co-authors. Others assert that it should be determined by degree of "intellectual contribution." Still others expect the writer of the first draft, who made the most important work effort, to be the first author and the senior co-author to be the last author.

Clearly, this inconsistency raises a great deal of confusion. Since a simple way to determine credit associated with the sequence of authors' names is still absent, the reader really does not know if the last author was the driving force or the least important in overall contribution to the work. A few emphasize that faculty should safeguard the rights of graduate students to publish the results of their research. The ICMJE requires that all authors (regardless of the number) take responsibility for the contents of the entire article, and not just for the task each may have performed.[11]

Notably, a few researchers have worked out their own methods of avoiding conflict over potential issues of credit. Michael Brown and Joseph Goldstein, awarded the 1985 Nobel Prize in medicine for their discovery of the mechanism of cholesterol metabolism, have worked together so amicably for years at the University of Texas Southwestern Medical School in Dallas that they earned the sobriquet Brownstein. They plan research jointly, publish together, and share the podium for lectures. Ernest

McCulloch and James Till at the Ontario Cancer Institute in Toronto took this a step further by alternating senior authorship on joint articles. In 2005, they received the Albert Lasker Award for Basic Medical Research for identifying the blood's stem cells and establishing their two main properties: self-renewal and differentiation into specialized cells that have limited life spans.[12] Barry Marshall, a gastroenterologist, and J. Robin Warren, a pathologist at the Royal Perth Hospital in Western Australia, chose to publish their landmark study in an unusual format: two separate letters in the *Lancet*. Warren's letter described the work on the stomach bacteria that he had conducted alone before collaboration with Marshall. Marshall's letter described their joint work. The two men published a joint report the following year indicating a breakthrough discovery: the bacterial cause of gastritis and of gastric and duodenal ulcers. In 2005, they shared the Nobel Prize in medicine.[13]

*Equitable and Impartial Peer Review and Openness to
Innovative and Paradigm-shifting Ideas*

I have covered the perils of peer review extensively in chapters 9 and 12. Questions regarding this system began arising in the 1970s and have become more insistent since the mid-1990s. Reviewers prefer submissions that mesh with their own perspective on how an issue should be conceptualized. Peers are, almost by definition, part of the established order and typically mired in traditional thinking. They are also human beings with their own agendas and priorities.

In 1997, Richard Smith, editor of the esteemed *British Medical Journal*, offered a trenchant critique: "The problem with peer review is that we have good evidence on its deficiencies and poor evidence on its benefits. We know that it is expensive, slow, prone to bias, open to abuse, possibly anti-innovatory, and unable to detect fraud."[14] As recently as 2006, an editorial in *Nature* asserted that peer review "provides only a minimal assurance of quality and that the public conception of peer review as a stamp of authentication is far from the truth."

In 1994, the General Accounting Office of the U.S. Congress studied the use of peer review in government scientific grants and found that reviewers often knew applicants and tended to give preferential treatment to those they knew.[15]

The Nobel Foundation has indicated the signposts of a creative mind that may lead to scientific breakthroughs: courage to think on entirely new lines, the ability to question established theories, and innovative combinations of insights from different fields. It may not be surprising, then, that an astonishingly large number of future Nobel laureates encountered resistance on the part of scientific journal editors.[16]

There is a lesson to be learned from experiences before the U.S. government began its massive research funding. The Rockefeller Foundation, which was the most expansive supporter of basic research in the United States, especially in biology between the two World Wars, relied on successful programs to find promising scientists. Its goal was to encourage imaginative thought and creative ideas. In Britain, too, the Medical Research Council believed in "picking the man, not the project" and nurturing successful results with progressive grants.

Whereas peer review was intended to protect the autonomy and self-governance of the sciences, it has become an agent for the defense of orthodoxy and a constraint on creativity. The review process should be modified to reduce the inherent bias toward prevailing concepts and to encourage mavericks and outsiders. Granting bodies should consider the qualities of creative people and the personae of the applicants as well as their proposals. The challenge is to have insight into whom to support before a track record exists, yet to discern frivolous proposals. A firm standard should be the degree to which a researcher's work threatens to disturb conventional beliefs.

As Henry Bauer, the author of *Scientific Literacy and the Myth of the Scientific Method*, has wryly observed: "One rarely noted aspect of peer review is that especially with the most brilliant ideas, the reviewers are less qualified than the authors of the research papers."[17] One approach to this problem is "science court"[18]—the establishment of a group of knowledgeable and competent scientists distributed over several fields with enough general proficiency to understand the detailed arguments of the field in question. This would provide relief from two major constraints of the traditional peer review system: unconscious bias and vested interests—intellectual, financial, commercial, status—that keep dogma entrenched and inhibit the exposure of innovative approaches.

Criteria for Determination of Credit and Acknowledgment of Contributions by Others

The most compelling historical example of the dilemma of attribution of credit revolves around who "discovered" America. Every schoolchild learns that it was Christopher Columbus. But the truth is, Columbus never knew what he had really found, clinging to the belief that the Caribbean Islands he kept bumping into were obscure pieces of Asia. "The notion that America was a continent, and that the body of water that separated his native home from the lands he was conquering was an ocean, separate from the waters of the east, just never occurred to him."[19] It fell to the Florentine navigator Amerigo Vespucci, who'd sailed along the coast of Brazil, to announce that the land across the sea was, in fact, a totally new continent.

Columbus may have discovered America, but Vespucci widely popularized his travels. Within a few years, German mapmakers published a new world map with the continent named: America, the feminine form of Vespucci's name.

This raises some fundamental questions: Does one have to know what one has discovered to be recognized as a discoverer? And what are the main factors that determine the attribution of credit? Is it the one who first conceives an idea? The one who conceives it and perhaps announces it among contemporaries? Or is it the person who applies it to humanity? Is it the individual who has the creativity to originate the idea or is it the person who brings it to fruition?

The expressed intentions of the two most coveted prizes in science are surprisingly ambiguous. Alfred Nobel's will dedicates the prizes "to those who, during the preceding year, shall have conferred the greatest benefit on mankind." The Albert Lasker Basic Medical Research Award "honors scientists whose fundamental investigations have provided techniques, information, or concepts contributing to the elimination of major causes of disability and death."

The question may not be so much "Who was the first?" because this always implicitly carries with it the additional question, "Who was the first to make a *real* difference?"

Coursing through the history of science is the axiom "the credit goes to the man who convinces the world, not to the man to whom the idea first occurs." Exceptions certainly have occurred, but generally the scientific world reserves little reward for precursors or progenitors but values those who nurture a new field. The thoughts serve to underscore two further points: due acknowledgment of the contributions of others and tolerance for expressions of dissent.

Harold Varmus's Nobel Prize acceptance speech in 1989 has set the standard for acknowledging who has meaningfully influenced one's own work; it is a model of attribution, as he cited more than forty other researchers whose findings had led him and his colleague to the basic discoveries of cancer-inducing genes.

New York Times science writer Nicholas Wade takes the view that the Nobel and other prize committees distort the true pattern of discovery, which is often an interactive process in which many researchers contribute vital pieces to the problem. He proposes that the history of the discovery be laid out in a detailed monograph so that all receive due credit for their work, even if the prize money is given just to the major contributor.[20]

If acknowledgment of contributions by others is a requisite, so is tolerance of dissent, since no appeals process currently exists for awards. One example besides Damadian stands out. As Watson and Crick were working on the structure of DNA in Cambridge, lengthy painstaking experimental work by Erwin Chargaff at Columbia University provided critical raw data: namely, that DNA contains equal amounts of the substances A (adenine) and T (thymine), on the one hand, and equal amounts of G (guanine) and C (cytosine), on the other. Watson then worked out that the "rungs" holding the double helix are constructed by these four nitrogenous bases in this order: AT, GC, TA, GC, AT, AT, GC, CA, the formula varying according to whether one of the bases is situated on the left or on the right of the double helix. Coupled with the insight derived from Rosalind Franklin's work, Crick then built a three-dimensional model of DNA. Chargaff not only loudly complained to worldwide colleagues at his being overlooked for his contributions but also published acerbic comments about Watson and Crick.[21] However, it was clearly not he who provided the crucial interpretation of the data.

At a time when stem cell research, national health care reform, and genomics are in the news on a daily basis, the subject of scientific research is very much in the public mind. A great secret of science has been revealed regarding its fundamentally ego-driven competitive nature. It benefits us greatly to understand the true dynamics of the discovery process and the attribution of credit for many reasons: Because priority is cherished as a universal right deserving of proper recognition. Because we are affected so directly by scientific advances. Because failure to acknowledge the prospects of an advance is often costly and unproductive. Because we need to be sound in our judgment of the allocation of funding and resources. Because profound benefits and consequences to society may be at stake.

ACKNOWLEDGMENTS

In the course of research for this book, several university archives and special collections departments were indispensable treasure troves of material, and their heads were paragons of gracious cooperation. I am indebted to Thomas Frusciano and his assistant, Erica Gorder, at Rutgers, the State University of New Jersey in New Brunswick; Thomas Whitehead at Temple University in Philadelphia; and F. Jason Torre and Kristen Nyitray at the State University of New York at Stony Brook. Jack Termine, archivist at SUNY Downstate Medical Center, Brooklyn, was similarly helpful.

I am grateful to others who have given me the benefit of their experience and knowledge: Joshua Becker, Francis Bonner, Terry Button, Elof Carlson, Raymond Damadian, Joseph Frank, Leonard Freeman, Alfred Goldhaber, Michael Goldsmith, Eileen Gregory, Clare Grey, Donald Hollis, Waylon House, David Kramer, Ching-Ming Lai, Vivan Lee, Richard Macchia, Alexander Margulis, James Mattson, Lawrence Minkoff, Robert Root-Bernstein, Robert Rossi, Leon Saryan, Alvin Silverstein, Amanda Sozer, Don Vickers, Byron Waksman, John Throck Watson, Heidi Weissmann, and Arnold Wishnia.

Douglas Eveleigh, professor of microbiology at Rutgers University, provided illuminating insights into the Waksman-Schatz affair. Albert Schatz's wife, Vivian, not only shared his experiences and reactions but also recalled details as if they happened yesterday. Diane Wendt, curator of medicine and science at the Smithsonian Institution, uncovered useful photos of early streptomycin artifacts. Bridget Jameison provided supportive material from the University of Illinois at Champaign-Urbana. Professor Amit Prasad of the Department of Sociology at the University of Missouri-Columbia offered instructive perspectives on the early development of MRI. Tereza Yerimayan of the Armenian National Committee of America clarified several historical facts. Dr. Hans Ringertz of the Karolinska Hospital, Stockholm, Sweden, provided unique insight into the deliberations of the Nobel Academy. Professor David Wright of Michigan State University offered guidance regarding aspects of research integrity and misconduct.

Many thanks as well to the research librarians at the Wellcome Institute for the History of Science in London, Cambridge University, the New York Academy of Medicine, and my local Emma S. Clark Memorial Library. Most of all in this regard, I am indebted to Colleen Kenefick, senior librarian at the Health Sciences Center at SUNY Stony Brook, who not only tracked the appropriate sources to all my queries but pointed me to other rewarding sources.

Susan Golant was indispensable in helping to shape the book's contents. My agent, Joëlle Delbourgo, placed me in the highly capable hands of my editor, Luba Ostashevsky, and her team at Palgrave Macmillan. My gratitude to my assistant, Alice Jimenez, for her constant dependability and selfless cooperation.

NOTES

Introduction

1. I am indebted to Roger Rosenblatt, *The Man In the Water: Essays and Stories* (New York: Random House, 1994), xv, for this anecdote.
2. H. David Politzer, "The Dilemma of Attribution," Nobel lecture, December 8, 2004, in *Les Prix Nobel, The Nobel Prizes 2004*, ed. Tore Frängsmyr (Stockholm: Nobel Foundation, 2005).
3. Carl Sagan, *Broca's Brain: Reflections on the Romance of Science* (New York: Random House, 1979), 18.
4. G. Mendel, "Versuche über pflanzen-hybriden. Verh. Naturf. Ver. Abhandlungen," *Brünn* 4 (1866): 3–47.
5. Hugo Iltis, *Life of Mendel* (New York: W.W. Norton, 1932), 282.
6. P. Rous, "A sarcoma of the fowl transmissible by an agent separable from the tumor cells," *Journal of Experimental Medicine* 13 (1911): 397–411.

Chapter 1: Stolen Credit

1. Michael Bliss, *The Discovery of Insulin* (Toronto, Ontario: McClelland and Stewart, 1982).
2. Richard Carter, *Breakthrough: The Saga of Jonas Salk* (New York: Trident, 1966), 1.
3. David M. Oshinsky, *Polio, An American Story* (Oxford: Oxford University Press, 2005), 175–176.
4. Ibid., 277–278.
5. "Win, Place and Show," editorial, *New York Times*, June 10, 1981.
6. H. P. Loomer, J. C. Saunders, and N. S. Kline, "A clinical and pharmacodynamic evaluation of iproniazid as a psychic energizer," *Psychiatric Research Report American Psychiatric Association* 8 (1958): 129–141.
7. Nathan S. Kline, "Clinical experience with iproniazid (marsilid)," *Journal of Clinical Experimental Psychopathology* 19 (2), suppl. 1 (1958): 72–78.
8. Author interview with Heidi Weissman, December 21, 2009.
9. Ibid.
10. Author interview with Leonard Freeman, July 28, 2011.
11. R. Grossman, "In academe, the serfs are toppling the lords," *Chicago Tribune*, August 24, 1997, C1.
12. Corynne McSherry, *Who Owns Academic Work: Battling for Control of Intellectual Property* (Cambridge, MA: Harvard University Press, 2011).

13. Stephen Brill, "When The Government Goes Judge Shopping," *The American Lawyer* (November 1988): 30.

14. "Aide Accuses Top Doctor of Plagiarism," *New York Times,* March 17, 1988.

15. Author Interview with Leonard Freeman, July 28, 2011.

16. United States Court of Appeals for the Second Circuit, *Heidi S. Weissman, M.D. v. Leonard M. Freeman, M.D.,* docket nos. 88–7435, 88–7465, argued October 20, 1988, decided February 23, 1989.

17. Ron Kaufman, "After 5 Years, Heated Controversy Persists in Science Copyright Case," *The Scientist* 6 (18) (September 14, 1992): 1, 4, 5, 10.

18. The National Coalition for Universities in Public Interest, "Are Montefiore Hospital and The Albert Einstein College of Medicine Protecting a Plagiarist?" *New York Times,* May 13, 1990, E19.

19. "Are Scientific Misconduct and Conflicts of Interest Hazardous to our Health?" Nineteenth Report by the Committee on Government Operations together with Dissenting and Additional Views, Washington, D.C., U.S. Government Printing Office, House Report 101–688.

20. "College Will Pay $900,000 to End Sex-Discrimination Suit," *Los Angeles Times*, March 18, 1994.

21. Author interview with Heidi Weissman, April 27, 2010.

22. R. K. Merton, "Priorities in Scientific Discovery: A Chapter in the Sociology of Science," *American Sociological Review* 22 (6) (December 1957): 635–659.

23. Eugene Strauss, *Rosalyn Yalow: Nobel Laureate* (Cambridge, MA: Perseus Books, 1998), 107–108.

24. S. Ramón y Cajal, "Recollections of my life," *Memoirs of the American Philosophical Society* 18(1) (1937): 546–547.

25. F. Crick, quoted in H. Zuckerman, *Scientific Elite: Nobel Laureates in the United States* (New York: The Free Press, 1977), 224.

26. Quoted in Mitchell Wilson, "How Nobel Prizewinners Get That Way," *The Atlantic,* December 1969.

27. R. K. Merton, "Priorities in Scientific Discovery: A Chapter in the Sociology of Science," *American Sociological Review* 22 (6) (December 1957): 635–659.

Chapter 2: The Art of Science

1. R. M. Wilson, *The Beloved Physician: Sir James Mackenzie* (New York: Macmillan, 1926), 177.

2. Rupert Lee, *The Eureka! Moment: 100 Key Scientific Discoveries of the 20th Century* (New York: Routledge, 2002).

3. J. D. North, *Isaac Newton* (Oxford: Oxford University Press, 1967).

4. Anthony Storr, *The Dynamics of Creation* (New York: Ballantine, 1993), 311–313.

5. Rollo May, *The Courage to Create* (New York: W.W. Norton & Company, 1994), 63.

6. A. Copland, quoted in Storr, *The Dynamics of Creation.*

7. C. P. Snow, "The Two Cultures," New Statesman and Society, October 6, 1956, 413–414, presented three years later as the Rede Lecture at Cambridge University.

8. C. Mooney and S. Kirshenbaum, *Unscientific America: How Scientific Illiteracy Threatens our Future* (New York: Basic Books, 2009).

9. W. H. Auden, *The Dyer's Hand and Other Essays. The Poet and the City* (New York: Random House, 1948, 1962), 81.

10. S. Chandrasekhar, *Truth and Beauty: Aesthetics and Motivations in Science* (Chicago: Chicago University Press, 1987).

11. Anne Fadiman, *Ex Libris* (New York: Farrar, Straus and Giroux, 1998), 91–92.

12. Isaac Asimov, *Biographical Encyclopedia of Science and Technology,* 2nd rev. ed. (New York: Doubleday, 1982), 415.

13. Mitchell Wilson, "How Nobel Prizewinners Get that Way," *Atlantic Monthly* 224 (1970): 69–74.

14. Storr, *The Dynamics of Creation*; Jacob Bronowski, *The Origins of Knowledge and Imagination* (New Haven, CT: Yale University Press, 1978); Arthur Koestler, *The Art of Creation* (London: Arkana, 1989), 120; Leonard Schlain, *Art and Physics: Parallel Visions in Space, Time and Light* (New York: Morrow, 1991); David Bohm, *On Creativity,* ed. Lee Nicholl (London: Routledge, 1998), 7, 15.

15. J. Michael Bishop, *How to Win the Nobel Prize* (Cambridge, MA: Harvard University Press, 2003), 54.

16. A. Appel, "An Interview with Nabokov," *Wisconsin Studies in Contemporary Literature* 8 (Spring 1967): 140–141.

17. Faraday, quoted in John Tyndall, *Faraday as a Discoverer* (London: Longmans, Green, 1868), 77–78.

18. I am indebted to Thomas G. West, *In the Mind's Eye* (Buffalo, NY: Prometheus Books, 1991), 9, for directing my attention to this episode.

19. L. Edson, "Two Men in Search of the Quark," *New York Times Magazine*, October 8, 1967, 64.

20. S. E. Luria, foreword to J. Bronowski, *The Origins of Knowledge and Imagination* (New Haven, CT: Yale University Press, 1978), x.

21. P. Buckley, and F. D. Peats, eds., *A Question of Physics: Conversations in Physics and Biology* (New York: Routledge and Kegan Paul, 1979), 129.

22. N. S. Kline, "Monamine oxidase inhibitors: An unfinished picaresque tale," in *Discoveries in Biological Psychiatry*, edited by Frank A. Ayd, Jr., and Barry Blackwell (Philadelphia: J.B. Lippincott Company, 1970), 195.

23. G. Edelman quoted in *Passionate Minds: The Inner World of Scientists*, edited by L. Wolpert and A. Richards (Oxford: Oxford University Press, 1997), 137.

24. A. H. Maslow, *Motivation and Personality,* 2nd ed. (New York: Harper & Row, 1970), 154.

25. G. Edelman quoted by S. Levy, "Annals of Science: Dr. Edelman's Brain," *The New Yorker* 70 (11) (1994).

26. Elkhonen Goldberg, *The Wisdom Paradox: How Your Mind Can Grow Stronger as Your Brain Grows Older* (New York: Gotham Books, 2005).

27. Timothy D. Wilson, *Strangers to Ourselves: Discovering the Adaptive Unconscious* (Cambridge, MA: Harvard University Press, 2002).

28. R. Rauschenberg quoted in David Markson, *This Is Not a Novel* (Washington, DC: Counterprint, 2001), 17.

Chapter 3: Staking the Claim

1. Sir Isaac Newton letter to Robert Hooke, February 5, 1676, in A. Rupert Hall, *Isaac Newton: Adventures in Thought* (Cambridge: Cambridge University Press, 1992).

2. Francis Darwin, ed., *The Life and Letters of Charles Darwin* (New York: Appelton, 1925), 426–427.

3. Ibid., 473.

4. Ibid., 474–475.

5. R. F. Furchgott, "A Research Trail Over Half a Century," *Annual Review of Pharmocological Toxicology* 35 (1995): 1–27.

6. Morton A. Meyers, *Happy Accidents: Serendipity in Modern Medical Breakthroughs* (New York: Arcade Publishers, 2007), 213–219.
7. R. F. Furchgott, D. Davidson, and C. I. Lim, "Conditions which determine whether muscarinic agonists contract or relax rabbit aortic rings and strips," *Blood Vessels* 16 (1979): 213–214.
8. Nicholas Wade, "Triumph of an 'Unworkable' Idea," *New York Times,* October 16, 1984, A30.
9. G. Köhler and C. Milstein, "Continuous cultures of fused cells secreting antibodies of predefined specificity," *Nature* 256 (1975): 495–497.
10. Nicholas Wade, obituary, "Georges Köhler, 48, Medicine Nobel Winner," *New York, Times,* March 4, 1995, 26.
11. Meyers, *Happy Accidents,* 213–219.
12. Nicholas Wade, *The Nobel Duel: Two Scientists' 21–Year Race to Win the World's Most Coveted Research Prize* (New York: Anchor Press/Doubleday, 1981).
13. Robert Gallo, *Virus Hunting: AIDS, Cancer and the Human Retrovirus, a Story of Scientific Discovery* (New York: Basic Books, 1991), 7.
14. "A Viral Competition over AIDS," *New York Times,* April 26, 1984, 22.
15. John Crewdson, *Science Fictions: A Scientific Mystery, a Massive Cover-Up, and the Dark Legacy of Robert Gallo* (Boston: Little, Brown and Company, 2002).
16. Robert C. Gallo and Luc Montaigner, "The discovery of HIV as the cause of AIDS," *The New England Journal of Medicine* 349 (December 11, 2003): 2283–2285.
17. Sinclair Lewis, *Arrowsmith* (New York: Harcourt, Brace Jovanovich, 1925).
18. Paul De Kruif, *The Microbe Hunters* (New York: Harcourt, Brace Jovanovich, 1926).
19. James Watson, *The Double Helix* (New York: Atheneum, 1968).
20. Brenda Maddox, *Rosalind Franklin: The Dark Lady of DNA* (New York: HarperCollins, 2002).
21. Gary Taubes, *Nobel Dreams: Power, Deceit and the Ultimate Experiment* (New York: Random House, 1986).
22. Allegra Goodman, *Intuition* (New York: Dial Press, 2006).

Chapter 4: The Dark Side of Science

1. J. L. Heilbron, "Creativity and Big Science," *Physics Today* (November 1992): 42–47.
2. D. Fanelli, "How many scientists fabricate and falsify research? A systematic review and meta-analysis of survey data," *PLoS ONE* 4 (5): e5738. Doi: 10.1371/journal pone.0005738.
3. William Broad and Nicholas Wade, *Betrayers of the Truth* (New York: Simon and Schuster, 1982).
4. Horace Freeland Judson, *The Great Betrayal: Fraud in Science* (Orlando, FL: Harcourt, 2004).
5. Salvatore Luria, "What makes a scientist cheat," *Prism* (May 15–18, 1975): 44, reprinted in J. Beckwith and T. Silhavy, *The Power of Bacterial Genetics* (Cold Spring Harbor, NY: Cold Spring Harbor Laboratory Press, 1992).
6. David Goodstein, "Scientific fraud," *American Scholar* 60 (1991): 505–515.
7. Robert K. Merton, "The Matthew effect in science," *Science* 159 (1968): 56–63.
8. Sidney Brenner, *My Life in Science* (London: Science Archive Limited, 2001).
9. Carl Djerassi, *Cantor's Dilemma* (New York: Doubleday, 1989), 113.
10. Anne Sayre, *Rosalind Franklin and DNA* (New York: W.W. Norton, 1975), 212, fn. 21 and 214, fn. 21.

11. D. Baltimore, "RNA-dependent DNA polymerase in virions of RNA tumor viruses," *Nature* 226 (1970): 1209–1211.

12. H. M. Temin and S. Mizutani, "RNA-dependent DNA polymerase in virions of rous sarcoma virus," *Nature* 226 (1970): 1211–1233.

13. Barbara J. Culliton, "The Sloan–Kettering Affair (II): An Uneasy Resolution," *Science* 184 (4142) (June 14, 1974): 1154–1157.

14. Ibid.

15. Jane Brody, "Change of False Research Data Stirs Cancer Scientists at Sloan-Kettering," *New York Times*, April 18, 1974, 20.

16. N. J. Steneck, "Fostering integrity in research: Definitions, current knowledge, and future directions," *Science and Engineering Ethics* 12 (2006): 53–74.

17. N. Lynöe, L. Jacobsson, and E. Landen, "Fraud, misconduct or normal science in medical research—an experimental study of demarcation," *Journal of Medical Ethics* 25 (1999): 501–506; R. De Vries, M. S. Anderson, and B. C. Martinson, "Normal misbehavior: Scientists talk about the ethics of research," *Journal of Empirical Research on Human Research Ethics* 1 (2006): 43–50.

18. Laurence K. Altman, "For Science's Gatekeepers, a Credibility Gap," *New York Times*, May 2, 2006.

19. William J. Broad, "Harvard Delays in Reporting Fraud," *Science* 215 (3421) (January 29, 1982): 478–482.

20. L. D. Claxton, "Scientific authorship. Part I: A window into scientific fraud," *Mutation Research* 589 (2005): 17–30.

21. William J. Broad, "Imbroglio at Yale (I): Emergence of a Fraud," *Science* 210 (October 10, 1980): 38–41.

22. Laurence K. Altman, "Columbia's Medical Chief Resigns; Ex-Associate's Data Fraud at Issue," *New York Times*, August 9, 1980, 1; William J. Broad, "Imbroglio at Yale (II): A Top Job Lost," *Science* 210 (October 10, 1980): 171–173.

23. A. S. Brown and D. R. Murphy, "Cryptomnesia: delineating inadvertent plagiarism," *Journal of Experimental Psychology* 15 (1989): 432–442; R. I. Marsh and G. H. Bower, "Eliciting cryptomnesia unconscious plagiarism in a puzzle task," *Journal of Experimental Psychology: Learning, Memory, and Cognition* 19 (1993): 673–688.

24. Arthur Koestler, *The Act of Creation* (London: Arkana, 1989), 120.

25. David Bohm, *On Creativity*, ed. Lee Nichol (London and New York: Routledge, 1998).

26. R. Luft, "Reminiscences and Reflections," *Mount Sinai Journal of Medicine* 59 (2) (1992): 97–98.

27. Judy Sarasohn, *Science on Trial: The Whistle–Blower, the Accused, and the Nobel Laureate* (New York: St. Martin's Press, 1993).

28. Daniel J. Kevles, *The Baltimore Case: A Trial of Politics, Science, and Character* (New York: W.W. Norton, 1998), 387.

29. David Hamilton, "Baltimore Throws in the Towel," *Science* 252 (5007) (May 10, 1991): 768–770.

30. Fintan Steele, "Clearing of researcher in 'Baltimore affair' boosts demand for reform," *Nature* 381 (June 27, 1996): 719–720.

31. John E. Walsh, *Unraveling Piltdown: The Science Fraud of the Century and Its Solution* (New York: Random House, 1996).

Chapter 5: "Drop Everything!"

1. H. B. Woodruff, "A Soil Microbiologist's Odyssey," *Annual Review of Microbiology* 35 (1981): 1–28.

2. M. A. Meyers, *Happy Accidents: Serendipity in Modern Medical Breakthroughs* (New York: Arcade Publishers, 2007), 63–81.

3. Lincoln Library of Essential Information, *Immigration to the United States for Selected Years* (Baltimore, MD: Frontier Press, 1928), 2100.

4. D. S. Landes, "Symposium on The Role of Research Universities in Innovation, Social Mobility, and Quality of Life in the 20th Century," Association of American Universities Centennial Meeting, April 17, 2000, available at http://aau.edu/aau/Landes.html.

5. S. A. Waksman, *My Life with the Microbes* (New York: Simon and Schuster, 1954), 17.

6. L. Pasteur, and J. F. Joubert, "Charbon et Septicémie," *Comptes Rendus de l'Académie des Sciences* (Paris) 85 (1877): 101–151.

7. C. Garré, "Ueber Antagonisten Unter Den Bacterien," *Correspbl. Schweiz. Ärz.* 17 (1887): 385–392.

8. Papacostas and Gate, "Les Associations Microbiennes Leurs," in *Applications Thérapeutiques* (Paris: Doin, 1928).

9. Paul Vuillemin, "Antibiose et Symbiose," *C.R. Assoc. Franç, Avanc. Sci.*, pt. 2 (1889): 525–543.

10. Waksman, *My Life with the Microbes*, 120.

11. S. A. Waksman, *Principles of Soil Microbiology*, 2d ed. (Baltimore, MD: Williams and Wilkins, 1932).

12. S. A. Waksman, "The antibiotic era: A history of the antibiotics and of their role in the conquest of infectious diseases and in other fields of human endeavor," The Waksman Foundation of Japan, 1975, 10.

13. S. A. Waksman and W. C. Davison, *Enzymes: Properties, Distribution, Methods and Applications* (Baltimore, MD: Williams and Wilkins, 1926).

14. C. Rhines, "The persistence of avian tubercle bacilli in soil and in association with soil micro–organisms," *Journal of Bacteriology* 29 (1935): 299–311.

15. S. A. Waksman, *The Conquest of Tuberculosis* (Berkeley, CA: University of California Press, 1964), 104.

16. S. A. Waksman, "Tenth anniversary of the discovery of streptomycin, the first chemotherapeutic agent found to be effective against tuberculosis in humans," *The American Review of Tuberculosis* 70 (1954): 1–8.

17. Waksman, *My Life with the Microbes*, 11.

18. Ibid., 233.

19. Meyers, *Happy Accidents*.

20. C. Gram, "Uber die isolierte Färbung der Schizomyceten in Schnitt–und Trockenpräparaten," *Fortschr Med* 2 (1884): 1985–1989.

21. R. J. Dubos, "Bactericidal effect of an extract of a soil bacillus on Gram-positive cocci," *Proc Soc Exper Biol Med* 40 (1939): 311; R. D. Hotchkiss, "From Microbes to Medicine: Gramicidin, René Dubos, and the Rockefeller," in *Launching the Antibiotic Era: Personal Accounts of the Discovery and Use of the First Antibiotics*, edited by Carol L. Moberg and Zanvil A. Cohn (New York: The Rockefeller University Press, 1990).

22. Waksman, *My Life with the Microbes*, 225.

23. Waksman, *The Conquest of Tuberculosis*, 114.

24. Harry Gilroy, "Waksman and 10,000 Microbes," *New York Times Magazine*, November 2, 1952, 20.

25. S. A. Waksman and H. B. Woodruff, "Streptothricin, a new selective bacteriostatic and bactericidal agent, particularly active against gram–negative bacteria," *Proc Soc Exp Biol Med* 49 (1942): 207–210; H. B. Woodruff and J. W. Foster, "*In vitro* inhibition of Mycobacteria by streptothricin," *Proc Soc Exper Biol Med* 57 (1944): 88–89.

26. Waksman, *The Conquest of Tuberculosis*, 106.

Chapter 6: The Star Pupil

1. I. Auerbacher and A. Schatz, *Finding Dr. Schatz: The Discovery of Streptomycin and a Life It Saved* (New York: iUniverse, Inc., 2006), 12.
2. Author interview with Vivian Schatz, August 22, 2009.
3. S. A. Waksman, *My Life with the Microbes* (New York: Simon and Schuster, 1954), 65.
4. Ibid., 82–83.
5. I. Auerbacher and A. Schatz, *Finding Dr. Schatz*, 32.
6. A. Schatz, "My Experience in World War II," Oral History Archives, Schatz Archives.
7. Waksman, *My Life with the Microbes*, 186–188.
8. S. A. Waksman, and R. Curtis, "The actinomyces of the soil," *Soil Science* I (1916): 99–134.
9. D. Jones, H. J. Metzger, A. Schatz, and S. A. Waksman, "Control of gram–negative bacteria in experimental animals by streptomycin," *Science* 100 (1944): 103–105.
10. Frank Ryan, *The Forgotten Plague: How the Battle against Tuberculosis Was Won—and Lost* (Boston: Little, Brown, 1993), 218.
11. Veronique Mistiaen, "Time and the great healer," *Manchester Guardian, Weekend*, London, November 2, 2002, 61–67.
12. Letter from Feldman to Waksman, November 24, 1943, Waksman archives, Rutgers University, New Brunswick, NJ.
13. Author interview with Vivian Schatz, March 10, 2008.
14. A. Schatz, E. Bugie, and S. A. Waksman, "Streptomycin, a substance exhibiting antibiotic activity against gram-positive and gram-negative bacteria," *Proc Soc Exp Biol Med* 55 (1944): 66–69.
15. J. H. Comroe, "Pay dirt: The story of streptomycin. Part I: From Waksman to Waksman," *American Review of Respiratory Diseases* 117 (1978): 773–781; "Part II: Feldman and Hinshaw; Lehmann," *American Review of Respiratory Diseases* 117 (1978): 957–968.
16. Auerbacher and Schatz, *Finding Dr. Schatz*, 41.
17. A. Schatz and S. A. Waksman, "Effect of streptomycin and other antibiotic substances upon *Mycobacterium tuberculosis* and related organisms," *Proc Soc Exp Biol Med* 57 (1944): 244–248.
18. Letter from Feldman to Waksman, May 11, 1944, Waksman archives, Rutgers University, New Brunswick, NJ.
19. S. A. Waksman, E. Bugie, and A. Schatz, "Isolation of antibiotic substances from soil microorganisms, with special reference to streptothricin and streptomycin," *Proceedings of the Staff Meetings of the Mayo Clinic* 19 (1944): 537–548.
20. W. H. Feldman and H. C. Hinshaw, "Effects of streptomycin on experimental tuberculosis in guinea pigs: A preliminary report," *Proceedings of the Staff Meetings of the Mayo Clinic* 19 (1944): 593–599.
21. Brent Hoff and Carter Smith III, "Tuberculosis," in *Mapping Epidemics: A Historical Atlas of Disease* (New York: Watts/Grolier, 2000), 94.
22. A. Schatz, "Streptomycin, an antibiotic agent produced by Actinomyces griseus" (Ph.D. diss., Rutgers University, New Brunswick, NJ, July 1945).
23. Draft of an undated lengthy memorandum by Waksman entitled "Contribution of Albert Schatz to the Development of Streptomycin," probably prepared in the spring of 1950 for his forthcoming deposition. Waksman archives, Rutgers University, New Brunswick, NJ.
24. Letter from Schatz to Waksman, May 21, 1946, Waksman archives, Rutgers University, New Brunswick, NJ.

25. Auerbacher and Schatz, *Finding Dr. Schatz*, 35.

26. Morton A. Meyers, *Happy Accidents: Serendipity in Modern Medical Breakthroughs* (New York: Arcade Publishers, 2007), 333, fn. 24.

27. Letter from Waksman to Dr. R. A. Strong, May 31, 1946, Waksman archives, Rutgers University, New Brunswick, NJ.

28. "The Healing Soil," *Time,* November 7, 1949; K. Chowder, "How TB survived its own death to confront us again," *Smithsonian* 23 (1992): 180–194; M. Smolen, "A Nobel Quest," *Rutgers Magazine* 71 (1992): 43–45; Ryan, *The Forgotten Plague*; Julie Fenster, *Mavericks, Miracles, and Medicine* (New York: Carroll and Graf Publishers, 2003).

29. Waksman's deposition in pre-trial hearing, May 1, 1950, p. 119, Schatz archives, Temple University, Philadelphia, PA.

30. Letter from Schatz to Waksman, January 31, 1948, Waksman archives, Rutgers University, New Brunswick, NJ.

31. Letter from Schatz to Waksman, November 29, 1948, Waksman archives, Rutgers University, New Brunswick, NJ.

32. Schatz's unpublished musings after receiving Waksman's letter of February 8, 1949, Schatz archives, Temple University, Philadelphia, PA.

33. Letter from Waksman to Schatz, January 12, 1949, Schatz archives, Temple University, Philadelphia, PA.

34. Letter from Schatz to Waksman, January 22, 1949, Waksman archives, Rutgers University, New Brunswick, NJ.

35. Waksman, *My Life with the Microbes*, 282.

36. Letter from Waksman to Schatz, January 28, 1949, and February 8, 1949, Waksman archives, Rutgers University, New Brunswick, NJ.

37. Letter from Waksman to Schatz, January 29, 1949, Schatz archives, Temple University, Philadelphia, PA.

38. Ibid.

39. Ibid.

40. Letter from Russell Watson to Schatz, February 8, 1949, Schatz archives, Temple University, Philadelphia, PA.

41. Author interview with Vivian Schatz, August 22, 2009.

42. Letter from M. D. Bromberg & Associates, Inc. to Dr. Elizabeth S. Clark, June 22, 1949, Waksman archives, Rutgers University, New Brunswick, NJ.

43. Letter from Doris Jones to M. D. Bromberg & Associates, Inc., June 30, 1949, Schatz archives, Temple University, Philadelphia, PA.

44. "Streptomycin profit asked by ex-student," *New York Times*, March 11, 1950, 16.

45. Letter from Robert C. Clothier to Waksman, March 15, 1950, Waksman archives, Rutgers University, New Brunswick, NJ.

46. Letter from George W. Merck to Waksman, February 14, 1945, Schatz archives, Temple University, Philadelphia, PA.

47. W. H. Helfand, H. B. Woodruff, K. M. H. Coleman, and D. L. Cowan, "Wartime industrial production of penicillin in the United States," in *The History of Antibiotics: A Symposium,* ed. John Parascandola (Madison, WI: American Institute of the History of Pharmacy, 1980), 31–56.

48. "Streptomycin Pays," *Time* 53, May 16, 1949, 87.

49. Waksman, *My Life with the Microbes*, 216.

50. Ibid., 282, 338.

51. "Waksman Royalty Is Put at $350,000: Earnings from Streptomycin Bared as Trial of Suit by Former Student Nears," *New York Times*, April 29, 1950, 7.

52. "Attorney Says Waksman Kept $350,000 Streptomycin Profit," *Newark Star-Ledger*, April 29, 1950.

53. "Waksman Royalty Is Put at $350,000: Earnings from Streptomycin Bared as Trial of Suit by Former Student Nears," *New York Times*, April 29, 1950, 7.
54. "Rutgers Is Too Smart for Its Own Good," *Passaic Herald News*, May 1, 1950.
55. Schatz's unpublished musings after receiving Waksman's letter of February 8, 1949, Schatz archives, Temple University, Philadelphia, PA.
56. Waksman's deposition in pre-trial hearing, May 1, 1950, 67, Schatz archives, Temple University, Philadelphia, PA.
57. Letter from Starkey to Waksman, May 10, 1950, Waksman archives, Rutgers University, New Brunswick, NJ.
58. "Dr. Waksman wins acclaim in Europe," *New York Times*, August 12, 1950.
59. Waksman, *My Life with the Microbes*, 284.
60. Author interview with Vivian Schatz, March 10, 2008.
61. Letter from William Feldman to Waksman, February 13, 1951, Waksman archives, Rutgers University, New Brunswick, NJ.
62. Letter from Walton Geiger to Waksman, January 18, 1951, Waksman archives, Rutgers University, New Brunswick, NJ.
63. "Dr. Schatz Wins 3% of Royalty; Named Co-Finder of Streptomycin," *New York Times*, December 30, 1950.
64. "Strepto–settlement," *Time* 57, January 8, 1951, 32.
65. The Streptomycin Litigation, Statement issued to the faculty by R. C. Clothier, President RREF, December 29, 1950, Waksman archives, Rutgers University, New Brunswick, NJ.
66. "Dr. Schatz Wins 3% of Royalty; Named Co–Finder of Streptomycin," *New York Times*, December 30, 1950.
67. Author interview with Vivian Schatz, August 29, 2008.
68. Statement in Shulhan Arak, Yore De'a 242, 1, based upon the Talmudic passages: Sanherdin 23a, 85b, 87a.
69. Author interview with Vivian Schatz, August 22, 2009.
70. H. B. Woodruff, "A Soil Microbiologist's Odyssey," *Annu Rev Microbiol* 35 (1981): 1–28.
71. Herbert Lechvalier, "The Search for Antibiotics at Rutgers University," in *The History of Antibiotics: A Symposium*, ed. John Parascandola (Madison, WI: American Institute of the History of Pharmacy, 1980).
72. B. Feldman, *The Nobel Prize: A History of Genius, Controversy, and Prestige* (New York: Arcade Publishers, 2000), 276.
73. Letter from Sabin to Elmer S. Reinthaler, vice president of the National Agricultural College, November 10, 1952, Schatz archives, Temple University, Philadelphia, PA.
74. Letter from Professor Coran Liljestrand, secretary of the Nobel Committee for Medicine for the Caroline Institute, reply to Reinthaler, November 14, 1952, Schatz archives, Temple University, Philadelphia, PA.
75. Letter from Schatz to the King of Sweden, December 6, 1952, Schatz archives, Temple University, Philadelphia, PA.
76. Waksman, *My Life with the Microbes*, 285.
77. Author interview with Vivian Schatz, March 10, 2008.
78. A. Schatz, "Some personal reflections on the discovery of streptomycin," *Pakistan Dental Review* 15 (1965): 125–134.
79. Albert Schatz, "The true story of the discovery of streptomycin," *Actinomycetes* 4 (2) (1993): 27–39.
80. Milton Wainwright, "Streptomycin: Discovery and Resultant Controversy," *History and Philosophy of the Life Sciences* 13 (1991): 97–124.
81. Author interview with Vivian Schatz, August 22, 2009.

82. Milton Wainwright, *Miracle Cure: The Story of Antibiotics* (Oxford, England: Basel Blackwell, 1990); Ryan, *The Forgotten Plague*; Feldman, *The Nobel Prize*.

83. V. Mistiaen, "Time and the Great Healer," *Manchester Guardian, Weekend,* London, November 2, 2002, 61–67; Peter A. Lawrence, "Rank injustice," *Nature* 415 (2002): 835–836.

84. Auerbacher and Schatz, *Finding Dr. Schatz.*

Chapter 7: Shock Waves in Academia

1. Albert Schatz, "The true story of the discovery of streptomycin," *Actinomycetes* 4 (2) (1993): 27–39.

2. H. B. Woodruff, *Scientific Contributions of Selman A. Waksman* (New Brunswick, NJ: Rutgers University Press, 1968).

3. D. Jones, H. J. Metzger, A. Schatz, and S. A. Waksman, "Control of gram-negative bacteria in experimental animals by streptomycin," *Science* 100 (1994): 103–105.

4. William Kingston, "Streptomycin, *Schatz v. Waksman*, and the Balance of Credit for Discovery," *Journal of the History of Medicine and Allied Sciences* 59(3) (July 2004): 441–462.

5. Samuel Epstein and Beryl Williams, *Miracles from Microbes: The Road to Streptomycin* (New Brunswick, NJ: Rutgers University Press, 1946), 136.

6. Sam Epstein, "Streptomycin Background Material," 12, Waksman archives, Rutgers University, New Brunswick, NJ.

7. H. Christine Reilly, pre-trial deposition, fall 1950, 433, Schatz archives, Temple University, Philadelphia, PA.

8. Interview by Judah Ginsberg, "Selman Waksman and Antibiotics," American Chemical Society website, available at http://portal.acs.org/portal/acs/corg /content?_nfpb=true&_pagelabel=PP_Articlemap...

9. Selman A. Waksman, "Microbiology Takes the Stage," *Scientific Monthly* 79 (6) (December 1954): 358.

10. Disclosed in an interview by Frank Ryan with Hubert Lechavalier, who succeeded Waksman as chair in microbiology at Rutgers (Frank Ryan, *The Forgotten Plague: How the Battle against Tuberculosis Was Won—and Lost* [Boston: Little, Brown, 1993], 442 fn. 29). In 1948, Lechavalier isolated neomycin, which is still in use today as a topical antibacterial agent: S. Waksman and H. Lechavalier, "Neomycin, a new antibiotic active against streptomycin resistant bacteria, including tuberculosis organisms," *Science* 109 (March 25, 1949): 305–307.

11. Author interview with Byron Halsted Waksman, January 28, 2008.

12. Selman A. Waksman, *The Conquest of Tuberculosis* (Berkeley: University of California Press, 1964).

13. Lee B. Reichan and Janice Hopkins Tanne, *Time Bomb: Global Epidemic of Multi-Drug-Resistant Tuberculosis* (New York: McGraw-Hill, 2002).

14. Clifton E. Barry III and Maija S. Cheung, "New Tactics Against Tuberculosis," *Scientific America* 300 (3) (March 2009): 62–69.

Chapter 8: "This Shameful Wrong Must Be Righted!"

1. Author interview with Raymond Damadian, May 1, 2008.

2. Author interview with Raymond Damadian, September 19, 2008.

3. James Mattson and Merrill Simon, *The Pioneers of NMR and Magnetic Resonance in Medicine: The Story of MRI* (Ramat Gan, Israel: bar-Ilan University Press. Published in the U.S.A. by Dean Books Co., Jericho, NY University Press, 1996).

4. Caroline Overington, "The man who did not win," *The Sydney Morning Herald*, October 17, 2003.
5. Earl Lane, "Shares prize for MRI innovation," *Newsday*, October 7, 2003.
6. Warren S. Strugatch, "L.I. @ Work: A Doctor's Quixotic Quest for a Nobel Prize," *The New York Times*, December 7, 2003.
7. Ibid.
8. Author interview with Raymond Damadian, September 19, 2008.
9. R. Damadian, "Tumor detection by nuclear magnetic resonance," *Science* 171 (1971): 1151–1153.
10. Paul C. Lauterbur, "All Science is Interdisciplinary—from Magnetic Moments to Molecules to Men," Nobel lecture, in *Les Prix Nobel. The Nobel Prizes 2003*, ed. Tore Frängsmyr (Stockholm: Nobel Foundation, 2004), pp. 245–51.
11. Author interview with Leon Saryan, September 22, 2008.
12. Author interview with Don Vickers, November 14, 2008.
13. Ibid.
14. Cited by Amit Prasad in "The (Amorphous) Anatomy of Invention: The Case of Magnetic Resonance Imaging (MRI)," *Social Studies of Science* 37 (2007): 533–560.
15. Author interview with Paul Lauterbur, June 29, 1994.
16. Paul C. Lauterbur, "One Path out of Many—How MRI Actually Began," in *Encyclopedia of Nuclear Magnetic Resonance, Vol. 1*, ed. David M. Grant and Robin K. Harris (New York: John Wiley & Sons, 1996), 445–449.
17. Ibid.
18. Prasad, "The (Amorphous) Anatomy of Invention."
19. Author interview with Don Vickers, November 14, 2008.

Chapter 9: The Race Is On

1. Author interview with Waylon House, October 27, 2008.
2. Donald P. Hollis, *Abusing Cancer Science: The Truth About NMR and Cancer* (self-published) (Chehalis, WA: The Strawberry Fields Press, 1987), 148. According to a letter from Hollis responding to a written query from me, the referees' letters were given to Hollis by Lauterbur.
3. Ibid.
4. P. C. Lauterbur, "Image formation by induced local interactions: Examples employing nuclear magnetic resonance," *Nature* 242 (1973): 190–191.
5. Paul C. Lauterbur, "One Path Out of Many—How MRI Actually Began," in *Encyclopedia of Nuclear Magnetic Resonance, Vol. 1*, ed. David M. Grant and Robin K. Harris (New York: John Wiley & Sons, 1996), 445–449.
6. Ibid., 447.
7. G. Edelman, quoted by S. Levy, "Annals of Science: Dr. Edelman's brain," *The New Yorker* 70 (11) (1994).
8. C. G. Fry, "The Nobel Prize in Medicine for Magnetic Resonance Imaging," *J Chem Educ* 81 (2004): 922–932.
9. Richard Ernst quoted in David M. Grant and Robin K. Harris, eds., *Encyclopedia of Nuclear Magnetic Resonance, Vol 1: Historical Perspectives* (New York: John Wiley & Sons, 1996), 93.
10. R. Smith, "Peer Review: Reform or Revolution?" *British Medical Journal* 315 (1997): 759–760.
11. Lauterbur, "One Path Out of Many—How MRI Actually Began," 447.
12. Quote from Lauterbur's grant application citing Damadian, as cited by Amit Prasad in "The (Amorphous) Anatomy of Invention: The Case of Magnetic Resonance Imaging (MRI)," *Social Studies of Science* 37 (2007): 533–560.

13. D. P. Hollis and L. A. Saryan, "A nuclear magnetic resonance study of water in two Morris hepatomas," *Johns Hopkins Medical Journal* 131 (6) (1972): 441–444.

14. Author interview with Leon Saryan, September 22, 2008.

15. Author interview with Waylon House, October 27, 2008.

16. Paul C. Lauterbur, "All Science is Interdisciplinary—from Magnetic Moments to Molecules to Men," Nobel lecture, in *Les Prix Nobel.* The Nobel Prizes 2003, ed. Tore Frängsmyr (Stockholm: Nobel Foundation, 2004), 245–51.

17. J. D. Eggleston, L. A. Saryan, D. P. Hollis, "Nuclear magnetic resonance investigation of human neoplastic and abnormal non-neoplastic tissues," *Cancer Res* 35 (1975): 1326–1332.

18. Author interview with Raymond Damadian, September 19, 2008.

19. R. Damadian, K. Zaner, D. Hor, and T. DiMaio, "Human Tumors Detected by Nuclear Magnetic Resonance," *Proc Nat Acad Sci* 71 (1974): 1471–1473.

20. Author interview with Raymond Damadian, September 19, 2008.

21. P. Mansfield and A. Mandsley, "Medical imaging by NMR," *Br. J. Radiol* 50 (1977): 188–194.

22. E. Raymond Andrew in *Encyclopedia of Nuclear Magnetic Resonance, Vol. 3*, ed. David M. Grant and Robin K. Harris (New York: John Wiley & Sons, 1996), 2467.

23. R. Damadian, L. Minkoff, M. Goldsmith, M. Stanford, and J. Koutcher, "Field focusing nuclear magnetic resonance (FONAR): Visualization of a tumor in a live animal," *Science* 194 (1976): 1430–1432.

24. Author interview with Lawrence Minkoff, January 21, 2009.

25. Author interview with Joseph Frank, August 26, 2008.

26. Sonny Kleinfield, *A Machine Called Indomitable* (New York: Times Books, 1985), 197.

27. Warren S. Strugatch, "L.I. @ Work: A Doctor's Quixotic Quest for a Nobel Prize," *New York Times*, December 7, 2003.

28. Author interview with Michael Goldsmith, September 9, 2008.

29. Author interview with Raymond Damadian, May 1, 2008.

30. Bettyann Holzmann Kevles, *Naked to the Bone* (Reading, MA: Helix Books, Addison-Wesley, 1997), 179.

31. E. Raymond Andrew in *Encyclopedia of Nuclear Magnetic Resonance, Vol. 3*, 2464.

32. Author interview with Lawrence Minkoff, January 21, 2009.

33. Author interview with Raymond Damadian, September 19, 2008.

34. Sir Peter Mansfield, "The Nobel Prize in Physiology or Medicine 2003," autobiography, available at http://nobelprize.org/nobel_prizes/medicine/laureates/2003/Mansfield.html.

35. Kleinfeld, *A Machine Called Indomitable*, 197.

36. Author interview with Michael Goldsmith, September 10, 2008.

37. Gordon Rattray Taylor, *The Science of Life* (New York: McGraw–Hill, 1963), 220.

38. R. Damadian, M. Goldsmith, and L. Minkoff, "NMR in Cancer. XVI FONAR Image of the Live Human Body," *Physiol Chem Phys* 9 (1977): 97–100.

39. Lawrence K. Altman, "New York Researcher Asserts Nuclear Magnetic Technique Can Detect Cancer, but Doubts Are Raised," *New York Times*, July 21, 1977.

40. Author interview with Raymond Damadian, September 19, 2008.

41. Ibid.

42. *Popular Science* 211 (6) (December 1977).

43. Author interview with Raymond Damadian, September 19, 2008.

44. R. Damadian, L. Minkoff, M. Goldsmith, and J. A. Koutcher, "Field-focusing nuclear magnetic resonance (FONAR): Formation of chemical scans in man," *Naturwissenschaften* 65 (1978): 250–252.

45. Author interview with Lawrence Minkoff, January 21, 2009.

Chapter 10: The Tipping Point

1. Author interview with Waylon House, January 21, 2009.
2. Ibid.
3. Ibid.
4. E. M. Purcell, Foreword, in C. L. Partain, R. R. Price, J. A. Patton, M. V. Kulikarni, and A. E. James Jr., *Magnetic Resonance Imaging*, 2d ed. (Philadelphia: W.B. Saunders Company, 1988), xxv–xxix.
5. H. Y. Carr, "Precession Techniques in Nuclear Magnetic Resonance" (Ph.D. diss., Harvard University, Cambridge, MA, 1952).
6. C.–N. Chen and D. I. Hoult, *Biomedical Magnetic Resonance Technology* (Bristol and New York: Adam Hilger, 1989), 38.
7. Purcell, Foreword, 1988, xxv–xxix.
8. Author interview with Raymond Damadian, September 19, 2008.
9. J. R. Singer, "Blood flow rates by NMR," *Science* 130 (1959): 1652–1653.
10. Author interview with Joseph Frank, August 26, 2008.
11. Author interview with David Kramer, October 22, 2008.
12. Alexander R. Margulis, "How NMR Was Started at the University of California, San Francisco (UCSF)," in *Encyclopedia of Nuclear Magnetic Resonance, Vol. 1: Historical Perspectives,* ed. David M. Grant and Robin K. Harris (New York: John Wiley & Sons, 1996), 484–485.
13. P. C. Lauterbur, "Cancer Detection by Nuclear Magnetic Resonance Zeugmatographic Imaging," *Cancer* 57 (1986): 1899–1904.
14. Waldo Hinshaw, in *Encyclopedia of Nuclear Magnetic Resonance, Vol. 1, Historical Perspectives,* ed. David M. Grant and Robin K. Harris (New York: John Wiley & Sons, 1996), 391.
15. David M. Grant, In *Encyclopedia of Nuclear Magnetic Resonance, Vol. 1, Historical Perspectives,* ed. David M. Grant and Robin K. Harris (New York: John Wiley & Sons, 1996), 120–121.
16. "Sir Peter Mansfield, The Nobel Prize in Physiology or Medicine 2003," autobiography, available at http://nobelprize.org.
17. P. Mansfield, I. L. Pykett, P. E. Morris, and R. Coupland, "Human whole body line-scanning imaging by NMR," *Br J Radiol* 51 (1978): 921–922.
18. W. S. Hinshaw, P. A. Bottomley, and G. N. Holland, "Radiographic thin-section imaging of the human wrist by nuclear magnetic resonance," *Nature* 270 (5639) (December 22–29, 1977): 722–723.
19. Sonny Kleinfield, *A Machine Called Indomitable* (New York: Times Books, 1985), 207.
20. Kleinfield, *A Machine Called Indomitable*, 1985.
21. Grant Fjermedal, "Inside Dr. Damadian's Magnet," *New York Times*, February 9, 1986.

Chapter 11: Obsession

1. F. W. Wehrli, "The origins and future of nuclear magnetic resonance imaging," *Physics Today* (June 1992): 38.
2. *FONAR Corp. v. General Electric Co.,* 107 F. 3d 1543 (Federal Circuit), certiorari denied, 522 U.S. 908 (1997).
3. Author interview with Raymond Damadian, June 12, 2008.
4. Paul Goldstein, *Intellectual Property* (New York: Portfolio, 2007), 38.
5. Ibid., 45.

6. Nicholas Varchaver, "The Patent King," *Fortune,* May 14, 2001, 202.

7. Paul C. Lauterbur, "One Path out of Many—How MRI Actually Began," in *Encyclopedia of Nuclear Magnetic Resonance, Vol. 1, Historical Perspectives,* ed. David M. Grant and Robin K. Harris (New York: John Wiley & Sons, 1996), 445–449.

8. Letter of recommendation by C. N. Yang to the Nobel committee dated June 1, 2000, copy provided by Raymond Damadian.

9. William Speed Weed, "The Way We Live Now: 12–14–03: Questions for Raymond Damadian; Scanscam?" *New York Times Magazine,* December 14, 2003.

10. Author interviews with Arnold Wishnia, May 13, 2008 and August 19, 2008.

11. Author interview with Leon Saryan, September 22, 2008.

12. M. Ruse, "The Nobel Prize in Medicine—Was there a religious factor in this year's (non) selection?" *Metanexus Online Journal,* March 16, 2004.

13. Mark O'Keefe, "Scientists are finding a friend in religion," available at http://www.catholiceducation.org.

14. Harvard conference on "The Quest for Knowledge, Truth and Values in Science and Religion," October 21–23, 2001, available at http://ecusa.anglican.org/19021_20678_ENG_HTM.htm.

15. William D. Phillips, "Does science make belief in God obsolete?" John Templeton Foundation, available at www.templeton.org/bigquestions.

16. Author interview with Raymond Damadian, September 19, 2008.

17. Horace Freeland Judson, "No Nobel Prize for Whining," *New York Times,* October 20, 2003, A17.

18. James Bernstein, Mark Harrington, and Earl Lane, "Medical Inventor Continues Lone Quest against Nobel Committee," *Newsday,* October 21, 2003.

19. J. Gore quoted in "MRI's inside story," *Economist* 369 (8353) (December 6, 2003).

20. Bernstein, Harrington, and Lane, "Medical Inventor Continues Lone Quest against Nobel Committee," *Newsday,* October 21, 2003.

21. Waldo Hinshaw, in *Encyclopedia of Nuclear Magnetic Resonance, Vol. 1, Historical Perspectives,* ed. David M. Grant and Robin K. Harris (New York: John Wiley & Sons, 1996), 391.

22. Paul C. Lauterbur, "All Science is Interdisciplinary—from Magnetic Moments to Molecules to Men," Nobel lecture, in *Les Prix Nobel.* The Nobel Prizes 2003, ed. Tore Frängsmyr (Stockholm: Nobel Foundation, 2004), 249.

Chapter 12: Picking the Winner

1. J. G. Crowther, *Science in Modern Society* (New York: Schocken Books, 1968), 363.

2. Morton A. Meyers, *Happy Accidents: Serendipity in Modern Medical Breakthroughs* (New York: Arcade Publishing, 2007), 38–91.

3. Thomas S. Kuhn, *The Structure of Scientific Revolutions* (Chicago: University of Chicago Press, 1962/1970).

4. A. Kantorovich and Y. Ne'eman, "Serendipity as a source of evolutionary progress in science," *Stud Hist Phil Sci* 20 (1989): 505–529.

5. Kuhn, *The Structure of Scientific Revolutions.*

6. D. J. DeSolla Price, *Little Science, Big Science* (New York: Columbia University Press, 1963).

7. John L. Heilbron, "Creativity and Big Science," *Physics Today* (November 1992): 42–47.

8. R. L. Weber and E. Mandoza, eds., *Random Walks in Science* (London, UK: Taylor & Francis, 2000).

9. Bruce G. Charlton, "Scientometric identification of elite revolutionary science research institutions by analysis of trends in Nobel prizes 1947–2006," *Medical Hypotheses* 68 (2007): 931–934.

10. S. Ramón y Cajal, *Advice for a Young Investigator* (Cambridge, MA: The MIT Press, 1999).

11. P. B. Medawar, *Advice to a Young Scientist* (New York: Basic Books, 1979).

12. P. Medawar, *Pluto's Republic* (Oxford, NY: Oxford University Press, 1982), 287.

13. Morton A. Meyers, "Back to the future," *AJR* 190 (2008): 561–564.

14. "Coping with peer rejection," *Nature* 425 (2003): 645.

15. Garrett Hardin, *Nature and Man's Fate* (New York: Holt, Richard and Winston, 1959).

16. J. C. McKinley, Jr., "Big Medical Research Prize Goes to 2 Pioneers in Genetics Work," *New York Times*, April 24, 2004, A18.

17. Meyers, *Happy Accidents,* 305–306.

18. J. M. Campanario, "Consolation for the scientist: Sometimes it is hard to publish papers that are later highly cited," *Social Studies of Science* 23 (1993): 342–362; J. M. Campanario, "Commentary on influential books and journal articles initially rejected because of negative referees' evaluations," *Science Communication* 16 (1995): 304–325; F. Godlee, "The ethics of peer review" in *Ethical Issues in Biomedical Publication,* ed. A. H. Jones and F. McLellan (Baltimore, MD: The Johns Hopkins University Press, 2000), 59–64.

19. J. Rösch, personal communication, January 9, 1995.

20. B. J. Marshall and J. R. Warren, "Unidentified curved bacilli in the stomach of patients with gastritis and peptic ulceration," *Lancet* 1 (1984): 1311–1315.

21. J. Folkman, personal communication, September 12, 1995.

22. Lawrence K. Altman, "3 Share Nobel in Medicine for a Breakthrough Gene Technique," *New York Times*, October 9, 2007, F3.

23. P. Strathern, *Mendeleyev's Dream: The Quest for the Elements* (New York: St. Martin's Press, 2001), 289.

24. Strathern, *Mendeleyev's Dream,* 289.

25. Oliver Sacks, "Brilliant Light," *The New Yorker*, December 20, 1999.

26. Robert Marc Friedman, *The Politics of Excellence. Behind the Nobel Prize in Science* (New York: W. H. Freeman Books, 2001), 33–34.

27. Burton Feldman, *The Nobel Prize: A History of Genius, Controversy, and Prestige* (New York: Arcade Publishing, 2000), 234–236.

28. Gerald Holton, *The Scientific Imagination: Case Studies* (Cambridge: Cambridge University Press, 1978), 64–67.

29. Robert Millikan, "The Isolation of an Ion, a Precision Measurement of Its Charge, and the Correction of Stoke's Law," *Science*, September 30, 1910.

30. Harvey Fletcher, "My work with Millikan on the oil-drop experiment," *Physics Today* 35 (June 1982): 43–47.

31. Ruth Lewin Sime, *Lise Meitner: A Life in Physics* (Berkeley, CA: University of California Press, 1996).

32. O. Hahn and F. Strassman, "Über den Nachweis und das Verhalten der bei der Bostrahlung des Urans mittels Neutronen entstehenden Erdalkalimetalle" (On the detection and characteristics of the alkaline earth metals formed by irradiation of uranium with neutrons), *Naturwissenschaften* 27 (1) (1939): 11–15.

33. Lise Meitner and O. R. Frisch, "Disintegration of Uranium by Neutrons: A New Type of Nuclear Reaction," *Nature* 143 (3615) (February 16, 1939): 239–240.

34. Friedman, *The Politics of Excellence,* 238–250.

35. Joshua Lederberg, "The Transformation of Genetics by DNA: An Anniversary Celebration of Avery, Macleod and McCarty," *Genetics* 136 (1994): 423.

36. Horace Freeland Judson, *The Eighth Day of Creation: The Makers of the Revolution in Biology* (New York: Simon & Schuster, 1979), 39.

37. Garland Allen, *Life Science in the Twentieth Century* (New York: Wiley, 1995).

38. Nobel Stiftelsen, *Nobel. The Man and His Prizes,* 3rd ed. (New York: American Elsevier, 1972).

39. Nicholas Wade, "Discovery of Pulsars: A Graduate Student's Story," *Science* 189 (August 1975): 358–364.

40. A. Hewish, S. J. Bell, J. D. H. Pilkington, P. F. Scott, and R. A. Collins, "Observation of a Rapidly Pulsating Radio Source," *Nature* 217 (February 24, 1968): 709–713.

41. Quoted in Horace F. Judson, *The Search for Solutions* (New York: Holt, Rinehart and Winston, 1980), 81–85.

42. Jocelyn Bell Burnell, "Little Green Men, White Dwarfs, or Pulsars?" *Cosmic Search* 1 (1). "Petit Four: After-dinner Speech," *Annals New York Academy of Sciences* 302 (1977): 685–689.

43. R. Freeman, E. Weinstein, E. Marineola et al, "Competition and Careers in Biosciences," *Science* 294 (2001): 2293–2295.

44. Quoted in "Education Life," *New York Times,* July 24, 2011.

45. J. L. Marx, "Lasker Award Stirs Controversy," *Science* 203 (January 26, 1979): 341.

46. Myron Pertz Glazer and Penina Migdal Glazer, *The Whistleblowers: Exposing Corruption in Government and Industry* (New York: Basic Books, 1989); C. Fred Alford, *Whistleblowers: Broken Lives and Organizational Power* (Ithaca: Cornell University Press, 2001), 134.

47. Terrance Miethe, *Whistleblowing at Work: Tough Choices in Exposing Fraud, Waste, and Abuse on the Job* (Boulder, CO: Westview Press, 1999).

48. Author interview with Heidi Weissman, April 27, 2010.

49. Alford, *Whistleblowers,* 120.

50. Ibid.

Epilogue

1. Quoted in profile of Harold Varmus by Natalie Angier, *New York Times*, November 21, 1993.

2. Henry H. Bauer, *Scientific Literacy and the Myth of the Scientific Method* (Chicago: University of Illinois Press, Urbana, 1994), 85.

3. Morton A. Meyers, *Happy Accidents: Serendipity in Modern Medical Breakthroughs* (New York: Arcade Publishing, 2007).

4. Author interview with Ernest McCulloch, January 25, 2006.

5. Lewis Thomas, foreword to *Natural Obsession: The Search for the Oncogene*, by Natalie Angier (Boston: Houghton Mifflin, 1988), xiii–xiv.

6. Bauer, *Scientific Literacy and the Myth of the Scientific Method*, 85.

7. D. Rennie, V. Yank, and L. Emanuel, "When authorship fails: A proposal to make contributors accountable," *JAMA* 278 (1997): 578–580.

8. E. Garfield. "More on the ethics of scientific publication: Abuses of authorship attributions and citation amnesia undermine the reward system of science," *Essays Inform Sci* 5 (1982): 621–626.

9. Mario Biagioli, "Rights or rewards? Changing frameworks of scientific authorship," in *Scientific Authorship: Credit and Intellectual Property in Science*, ed. Mario Biagioli and Peter Galison (New York: Routledge, 2003), 253–279.

10. L. D. Claxton, "Scientific authorship. Part 2. History, recurring issues, practices, and guidelines," *Mutation Research* 589 (2005): 31–45.

11. "Uniform requirements for manuscripts submitted to biomedical journals," *JAMA* 277 (1997): 928.

12. Meyers, *Happy Accidents*, 165–168.

13. Ibid., 99–116.

14. Richard Smith, "Peer review: Reform or revolution?" *British Medical Journal* 315 (1997): 759–760.

15. General Accounting Office, "Peer Review: Reforms Needed to Ensure Fairness in Federal Agency Grant Selection," June 24, 1994, GAO/PEMD-94-1.

16. Juan Miguel Campanario, "Consolation for the Scientist: Sometimes It Is Hard to Publish Papers That Are Later Highly Cited," *Social Studies of Science* 23 (1993): 342–362. J. M. Campanario, "Commentary on Influential Books and Journal Articles Initially Rejected Because of Negative Referees' Evaluations," *Science Communication* 16 (1995): 304–325. F. Godlee, "The Ethics of Peer Review," in *Ethical Issues in Biomedical Publication*, ed. Anne Hudson Jones and Faith McLellan (Baltimore, MD: Johns Hopkins University Press, 2000), 59–64.

17. Henry H. Bauer, *Scientific Literacy and the Myth of the Scientific Method*, 117.

18. Thomas Gold, "New Ideas in Science," *Journal of Scientific Exploration* 3 (1989): 103–112.

19. Simon Winchester, *Atlantic* (New York: HarperCollins, 2009), 88.

20. Nicholas Wade, "Eyes on the Prize," *New York Times*, April 30, 1995, SM24.

21. Erwin Chargaff, *Essays on Nucleic Acids* (Amsterdam: Elsevier, 1963); "A Quick Climb up Mount Olympus," *Science* 159 (1968): 1448–1449; Robert Olby, "The Path to the Double Helix," *Perspectives in Biology and Medicine* 19 (1976): 289–290.

SELECTED
BIBLIOGRAPHY

Alford, C. Fred. *Whistleblowers: Broken Lives and Organizational Power.* Ithaca: Cornell University Press, 2001.

Auerbacher, Inge and Albert Schatz. *Finding Dr. Schatz: The Discovery of Streptomycin and a Life It Saved.* New York: iUniverse, Inc., 2006.

Bauer, Henry. *Scientific Literacy and the Myth of the Scientific Method.* Urbana, IL: University of Illinois Press, 1994.

Beveridge, W. I. B. *The Art of Scientific Investigation.* New York: Vintage, 1950.

Bishop, J. Michael. *How to Win the Nobel Prize.* Cambridge, MA: Harvard University Press, 2003.

Bohm, David. *On Creativity.* Edited by Lee Nichol. London: Routledge, 1998.

Bryson, Bill, ed. *Seeing Further: The Story of Science, Discovery and the Genius of the Royal Society.* New York: William Morrow, Imprint of HarperCollins Publishers, 2010.

y Cajal, S. Ramón. *Advice for a Young Investigator.* Cambridge, MA: The MIT Press, 1999.

Crewdson, John. *Science Fictions: A Scientific Mystery, a Massive Cover-up, and the Dark Legacy of Robert Gallo.* Boston: Little, Brown and Company, 2002.

Djerassi, Carl. *Cantor's Dilemma.* New York: Doubleday, 1989.

———. *The Bourbaki Gambit.* New York: Penguin, 1994.

Feldman, B. *The Nobel Prize: A History of Genius, Controversy, and Prestige.* New York: Arcade, 2000.

Friedman, Robert Marc. *The Politics of Excellence: Behind the Nobel Prize in Science.* New York: W. H. Freeman, Times Books, Henry Holt and Company, 2001.

Goldberg, Elkhonen. *The Wisdom Paradox: How Your Mind Can Grow Stronger as Your Brain Grows Older.* New York: Gotham Books, 2005.

Goodman, Allegra. *Intuition.* New York: Dial Press, 2006.

Hixson, Joseph. *The Patchwork Mouse.* Garden City, NY: Anchor Press, 1976.

Judson, Horace Freeland. *The Great Betrayal: Fraud in Science.* Orlando, FL: Harcourt, 2004.

Kevles, Daniel J. *The Baltimore Case: A Trial of Politics, Science, and Character.* New York: W.W. Norton, 1998.

Kleinfield, Sonny. *A Machine Called Indomitable.* New York: Times Books, 1985.

Koestler, Arthur. *The Art of Creation.* London: Arkana, 1989.

Kuhn, Thomas S. *The Structure of Scientific Revolutions.* Chicago: University of Chicago Press, 1962/1970.

Mattson, James and Merrill Simon. *The Pioneers of NMR and Magnetic Resonance in Medicine: The Story of MRI.* Ramat Gan, Israel: Bar-Ilan University Press. Published in the U.S.A. by Dean Book Co. Jericho, NY: University Press, 1996.

May, Rollo. *The Courage to Create*. New York: W.W. Norton, 1975.

McCalman, Iain. *Darwin's Armada: Four Voyages and the Battle for the Theory of Evolution*. New York: W.W. Norton, 2009.

McSherry, Corynne. *Who Owns Academic Work? Battling for Control of Intellectual Property*. Cambridge, MA: Harvard University Press, 2001.

Medawar, P. B. *Advice to a Young Scientist*. The Alfred P. Sloan Foundation Series. New York: Basic Books, 1979.

Meyers, Morton A. *Happy Accidents: Serendipity in Modern Medical Breakthroughs*. New York: Arcade Publishing, 2007.

Norrby, Erling. *Nobel Prizes and Life Sciences*. New Jersey, Singapore: World Scientific, 2010.

Root-Bernstein, Robert and Michele. *Sparks of Genius: The Thirteen Thinking Tools of the World's Most Creative People*. Boston: Houghton Mifflin Company, 1999.

Ryan, Frank. *The Forgotten Plague: How the Battle against Tuberculosis Was Won—and Lost*. Boston: Little, Brown, 1993.

Sarasohn, Judy. *Science on Trial: The Whistle-Blower, the Accused, and the Nobel Laureate*. New York: St. Martin's Press, 1993.

Schlain, Leonard. *Art & Physics: Parallel Visions in Space, Time, and Light*. New York: William Morrow, 1991.

Sime, Ruth Lewin. *Lise Meitner: A Life in Physics*. Berkeley, CA: University of California Press, 1996.

Storr, Anthony. *The Dynamics of Creation*. New York: Atheneum, 1972.

Taubes, Gary. *Nobel Dreams: Power, Deceit, and the Ultimate Experiment*. New York: Random House, 1986.

Wade, Nicholas. *The Nobel Duel: Two Scientists' 21-Year Race to Win the World's Most Coveted Research Prize*. New York: Anchor Press/Doubleday, 1981.

Waksman, S. A. *The Conquest of Tuberculosis*. Berkeley, CA: University of California Press, 1964.

———. *My Life with the Microbes*. New York: Simon and Schuster, 1954.

Watson, James. *The Double Helix*. New York: Atheneum, 1968.

Wolpert, Lewis and Alison Richards. *Passionate Minds: The Inner World of Scientists*. Oxford: Oxford University Press, 1997.

Zuckerman, Harriet. *Scientific Elite: Nobel Laureates in the United States*. New York: The Free Press, 1977.

ILLUSTRATION CREDITS

DBC   Mattson, J. and M. Simon, editors of "The Pioneers of NMR and Magnetic Resonance in Medicine: The Story of MRI," Bar-Ilan University Press. Published in the U.S.A. by Dean Books Company, Jericho, New York, 1996.

N   Reproduced with permission from P. C. Lauterbur, "Image formation by induced local interactions. Examples employing nuclear magnetic resonance." *Nature* 242 (1973): 190–191.

RD   Courtesy of Raymond Damadian

SB   Special Collections and University Archives, State University of New York at Stony Brook

SI   Smithsonian Institution, Division of Medicine and Science, Washington, DC

VS   Courtesy of Vivian Schatz

WA   Waksman Archive, Special Collections and University Archives, Rutgers University, New Brunswick, NJ

WH   Courtesy of Waylon House

INDEX